A Synopsis of Craniofacial Growth

A Synopsis of Craniofacial Growth

DON M. RANLY D.D.S., Ph.D.

Professor and Chairman
Department of Pedodontics
Professor, Department of Physiology
University of Texas, Dental Branch
Houston, Texas

APPLETON-CENTURY-CROFTS/Norwalk, Connecticut

83 84 85 86 87 | 10 9 8 7 6 5 4 3

Prentice-Hall International, Inc., London
Prentice-Hall of Australia, Pty. Ltd., Sydney
Prentice-Hall of India Private Limited, New Delhi
Prentice-Hall of Japan, Inc., Tokyo
Prentice-Hall of Southeast Asia (Pte.) Ltd., Singapore
Whitehall Books Ltd., Wellington, New Zealand

Library of Congress Cataloging in Publication Data
Ranly, Don M   1938–
    A synopsis of craniofacial growth.

    Bibliography: p.
    Includes index.
    1. Skull—Growth.   2. Face—Growth.
3. Dentition.   I. Title. [DNLM: 1. Maxillofacial development.   2. Skull—Growth and
development. WU101.3 R211s]
QM535.R32   612.6'401'715   79-18423
ISBN 0-8385-8779-8

Text and cover design by Jane Breskin Zalben

Printed in the United States of America

This Book Is Dedicated to
Lili and Edmund,
and
All the Other Children of the World—
May Their Faces Grow Up with Smiles

Contents

Preface

This book is intended for upper-level dental students, graduate dentists in specialty programs, and practitioners in orthodontics, pedodontics, or family practice. It should also prove of value to allied professions where craniofacial growth is a concern.

Although designed as an introduction to the topic, the presentation presupposes a background of anatomy and some histology and biochemistry.

As an overview of a broad and complex subject, the content must necessarily be limited in scope and depth. The selection of subject matter was neither capricious nor exclusionary; omission of findings or concepts does not imply a value judgment. Simply, the abundance of source material greatly exceeded the space available.

The author has attempted to blend the central topics of craniofacial growth, both basic and applied, with the theoretical and speculative. It is hoped that such an approach will adequately cover the factual material, while permitting at the same time discussion of the uncertainties that remain in the field, the evolution of craniofacial thinking, and possible directions for the future.

The beginning student of craniofacial growth should find this volume a stepping-stone into a fascinating subject. For those already grounded in the fundamentals, it might serve as a useful review to tie together the disparate topics of the field.

Acknowledgments

The author would like to express his gratitude to the many people involved with this project. Typing assistance was unselfishly provided by Sheril McCue. Linda Moore offered a good shoulder and unflagging assistance. Karen Dale performed yeoman service in translating my thoughts into the written word. Dr. Jacob Geller and Dr. Dan West were generous of their time and constructive criticism. The graphics department of the university of Texas Dental Branch at Houston labored arduously to bring the illustrations to fruition.

Special thanks must go to Susan White for typing and editing the many drafts, and for the unwavering moral support, without which this book might never have been born.

And thanks to my family, colleagues, and friends who indulged me the time that was stolen from other responsibilities.

A Synopsis of Craniofacial Growth

· 1 ·

Introduction

Do not think that years leave us and find us the same!
Lord Lytton

The unfolding of craniofacial growth is a beautiful and complex phenomenon. One might describe it in superficial terms such as changes in proportion, lengthening or broadening of facial features, or age-related profile differences. Certainly these kinds of descriptions are not without meaning, for it is in their observation and appreciation that part of the growth and development of every human is apprised.

Lay people are instinctively aware of the relationship between time and growth in the face of the child, but if they were asked to compile the yearly changes, it is likely that a haphazard collection of data would result. The artist trained to do portraits could speak of bones and muscles beneath the skin, could discuss changes in profile and prominence of facial features, and could describe the effects of skin wrinkles and smile lines. But to the clinician concerned with the process of facial growth, such descriptive terms constitute only a fraction of the parameters needed. We should not be too eager, however, to discount the importance of this level of analysis, for ultimately only that which can be seen by the naked eye is projected to the rest of mankind. The orthodontist, for example, is vitally concerned with facial balance and harmony, because the smile and profile that he is able to mold into a patient's face is carried forth to meet the real world. And it is reasonable for us to speculate about how the success or failure of such treatment can affect the child's social and psychologic well-being.

But for any clinician to achieve the optimum facial features possible, he must think beyond external manifestations to the underlying changes in the bony foundation that brings them about. The student of facial growth must be interested in where growth occurs and what tissue and cells are contributing. He must also ask *why* questions, for which, at our current knowledge level, there are few answers. He must attempt to differentiate between normal and abnormal growth (sometimes no mean feat), and he must, if he is a therapist, understand what patterns can be altered and what techniques to do so are presently available. Finally, the serious student of facial growth should endeavor to understand the methodology and implications of growth prediction.

In the study of any complex subject, dissection of the whole into its parts facilitates the examination of the where, what, and why questions. In cell biology, for example, much more has been learned about the functions of the cell through analysis of cell organelles than study of intact cells themselves. Consider the wealth of information garnered from experimentation on isolated mitochondria, endoplasmic reticulum, nuclei, and even suborganelles such as chromosomes. In like manner, the study of facial growth has proceeded at every level imaginable (clinical to electron microscopic), from every discipline (surgical to philosophic), and with numerous objectives (therapeutic to theoretical). As a result, there is a plethora of available data, observations, and philosophies from which to choose in the preparation of a text on facial growth. I hope the method of presentation selected will embody the pertinent information from most of the contributing disciplines, combining enough facts with sufficient theory to proffer an overview of current thinking on growth of the head.

BONE—THE EMPHASIS

Bone, by virtue of its supporting function, as a result of its density (which allows roentgenographic visualization), and because of its contribution to facial form, has become the focal point of the majority of craniofacial research. Soft tissue unquestionably has major impact on bone and growth, but it is easier to measure the effect than the cause in this situation. The nature of bone encourages dimensional analysis either in vitro (on skulls) or in vivo (through the utilization of cephalometric x-rays). With ground sections, the footprints of time and change can be deciphered from the patterns of apposition and resorption.

In contrast, consider the difficulties that accrue to soft-tissue analysis. Calipers and rulers elucidate little when used on compressible, elastic materials, and the individual components of craniofacial soft tissues are patently not distinct enough for roentgenographic interpretation.

For these reasons and many others, the emphasis of this text will be on bone, both as tissue and as an organ.

CONTENTS

Chapter 2, *Mechanisms of Bone Growth,* will present the variety of ways new bone tissue is created. It will include the cells and tissues of matrix formation and calcification, describing their roles in the direct formation of bone as well as the indirect method utilizing cartilage precursors.

Chapter 3, *Principles of Bone Growth,* will explore the fascinating field concerned with the elucidation of those patterns or characteristics of bone growth and remodeling that are common to all bones, regardless of their anatomic locations. In formulating a few fundamental principles that govern the size and shape of bones, craniofacial investigators have narrowed the gap between biology and physics.

Chapter 4, *Description of Growth by Anatomic Divisions,* explains the benefits and rationale of analyzing the head by regions. Incorporating information from the previous two chapters, a systematic examination of growth in the cranium–cranial base complex, the middle face, and the mandible, will be presented.

Chapter 5, *Integration of Craniofacial Growth,* will reassemble the separate growth patterns of the anatomic divisions and meld them into an overall picture of craniofacial growth, emphasizing the interrelationships between the components.

Chapter 6, *Theories of Craniofacial Growth,* will explore the intriguing why questions of craniofacial biology. Interest in abstract matters signaled by such questions as "Why does the face grow downward and foreward?" or "Why must the eyeball be present for normal orbit development?" or "Why does a perverted swallowing habit result in a malocclusion?" is not really of an esoteric nature. Any clinician proposing therapy involving growth alterations should pay more than lip service to whys of craniofacial biology.

Chapter 7, *Growth Prediction,* will be devoted to the theoretical and clinical aspects of predicting facial growth and how this ability would be of immeasurable value to the clinician.

Chapter 8, *Development of the Dentition,* completes the description of the growth of the head by presenting the embryology of the tooth, the theories of eruption, and the periods characteristic of human dental development.

Chapter 9, *Principles of Cephalometric Analysis,* will present the basics of two fundamentally different cephalometric analyses that are used to evaluate normal and abnormal facial growth.

· 2 ·

Mechanisms of
Bone Growth

There's no art to find the mind's construction in the face.
Shakespeare, Macbeth

BONE AS A TISSUE

Introduction

The superstructure of the body is composed of bone and cartilage. Early in the development of each individual, cartilage is the dominant component; at maturity, bone easily predominates. The transition from one to the other is a complex phenomenon currently engaging the interest of scientists from a broad range of diciplines.

Histologically, the two tissues can readily be differentiated. Not only is bone calcified, but its internal architecture and its variety of cells are quite distinct. Biochemically, the differences between bone and cartilage are less marked but, nevertheless, significant. While the matrices of both tissues are composed of collagen and protein–carbohydrate complexes, variations in the chemical composition of these groups are sufficient to irrevocably cast the whole tissue as either bone or cartilage. Thus, bone is not simply mineral-impregnated cartilage with a revised internal form. Rather, the conversion from one to the other requires a complete turnover in matrix, as

5

well as change in cell types, mineral content, and organization.

To better understand the nature of the cartilage and bone and the transition from one to the other, a cursory overview of the molecular profile of these tissues will be presented.

Organic Matrix of Bone

Collagen

The major organic constituent of cartilage and bone is called collagen. It is essentially protein, save for a few important carbohydrate side groups. Collagen exists as highly oriented fibers that serve to provide strength for the connective tissues of the body.

In the electron microscope, the fibers are seen to be formed by the parallel alignment of microfibrils, which display a characteristic pattern of cross-striations, or bands. The most prominent cross-striations are about 680 Å apart and have been ascribed to the specific alignment of the basic molecular units of collagen, tropocollagens. These units have the dimension of a rod 15 Å in diameter and 3000 Å long.

The periodicity of the cross-striations is thought to be due to charged regions on the tropocollagen molecules, which, when registered with others in a precise array, stain in the characteristic 680 Å pattern (Fig. 2.1).

The tropocollagen molecule consists of three polypeptide chains of approximately 1000 amino acids each. These chains, individually and collectively, are coiled in unique, rigid, helical structures. Each of these polypeptide subunits is called an alpha chain. There are several genetically different collagens that are identified according to the amino acids sequence of the individual chains and their pattern of aggregation. For instance, the tropocollagen molecule may consist of three identical alpha chains and thus be designated as $[\alpha 1]_3$. If the tropocollagen consists of two identical alpha chains and another that differs in amino acid sequence, it is abbreviated as $[\alpha 1]_2 \alpha 2$. To show that the composition of the alpha 1 chain of $[\alpha 1]_3$ differs from the alpha 1 chain of $[\alpha 1]_2 \alpha 2$, roman numerals are added to the symbol, e.g., $[\alpha 1(I)]_3$ and $[\alpha 1(II)]_2 \alpha 2$.

It has now been determined that embryonic and adult bone contains tropocollagen, which has a chain distribution consisting of two $\alpha 1$-type I chains and one $\alpha 2$ chain. This grouping is abbreviated $[\alpha 1(I)]_2 \alpha 2$, or simply type I collagen. In embryonic and adult cartilage, the predominant tropocollagen consists of three identical chains, different in amino acid composition from that of bone and accordingly designated $[\alpha 1(II)]_3$ or type II collagen.

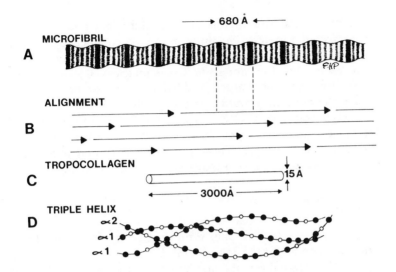

FIGURE 2.1   Diagrammatic representation of collagen structure. (A) The cross-striations of the microfibril with the regular repeat of 680 Å. (B) A two-dimensional view of the alignment of tropocollagen units forming the microfibril. (C) The essential dimensions of the rodlike unit tropocollagen. (D) The formation of tropocollagen by the aggregation of three polypeptides called alpha chains. Adapted from M.E. Grant and D.J. Prockop: The biosynthesis of collagen. New Eng. J Med 286:194, 1972.

Proteoglycans

The other major organic constituent of cartilage and bone consists of the protein–carbohydrate complexes formed by the covalent linkage of glycosaminoglycans (GAG) to protein cores. The GAGs (formerly known as acid mucopolysaccharides) are highly charged polyanions composed of repeating disaccharide units bearing either a sulfate or a carboxyl group or both. Eight of these complex molecules have been studied:—hyaluronate, chondroitin, chondroitin 4- and 6-sulfates, keratan sulfate, dermatan sulfate, heparin sulfate, and heparin—and each possess a different repeating disaccharide moiety. Most GAGs exist in vivo bound to core protein, radiating out in the fashion of a bristle brush.

These monomers, in turn, are linked together by hyaluronate, which reacts with GAG-free ends of protein cores. Huge complexes of proteoglycans, bound together by hyaluronate, are very important in the structural integrity of cartilage. Less is known about bone because of the experimental difficulties associated with the calcified matrix.

Hyaluronate, as an individual GAG, might have other functions in the extracellular matrix. The unusual physicochemical behavior of hyaluronate has suggested various roles for this huge molecule in initiating and stabilizing connective tissue matrices. Because of their immense size, negative charges, and extended form, the entangling overlap of the molecules could create a special environment with the capacity (1) to trap water, (2) to disrupt the diffusion of solutes, (3) to exclude macromolecules or particles from the meshwork, and (4) to create an osmotic pressure. These properties and their relationship to skeletogensis will be discussed later.

Cells of Bone

Although bone is a dense tissue, at no time, even after formation, is it static. The deposition during growth, the remodeling of development, and the internal transformation of maturation (as well as the control of mineral homeostasis) are made possible by the cells of bone. There are three classes of bone cells: the osteoblasts, the osteocytes, and the osteoclasts. The functions ascribed to each are generally accepted; the origin in some cases is still controversial.

The Osteoblast

This cell is responsible for the synthesis of the extracellular matrix of bone (the collagen and proteoglycans) and the initiation of calcification. The osteoblasts are generally incapable of cell division and must arise by modulations from precursor mesenchymal cells.

An active osteoblast is columnar in shape and is characterized by the following organelles: a single nucleus, an elaborate endoplasmic reticulum, a well-developed Golgi apparatus, and many mitochrondria. These are the earmarks of a cell actively engaged in synthesis.

The osteoblast is a surface cell found in the deep layer of the periosteum, around the trabeculae, or lining the canals of the haversian system. As they deposit the matrix, the osteoblasts retreat, maintaining for a time their positions on the surface. Invariably, the cells cease their unilateral secretion and envelop themselves in their own juices. Once osteoblasts become buried within the bone, they are called osteocytes.

The osteoblast manufactures the collagen on the ribosomes located along the endoplasmic reticulum. Straight-chain peptides of approximately 1000 amino acids and with a molecular weight of 115,000 to 125,000 are synthesized. By virtue of different amino acid sequences, two peptides are produced in a 2:1 stoichiometric relationship. Two $\alpha 1$ and one $\alpha 2$ peptides unite in a triple helical structure called procollagen, and these molecules accumulate within the cisternae of the reticulum. From here these molecules are transported to the Golgi apparatus, where they are presumably concentrated into vesicles. The material within these vesicles occasionally shows a 680 Å periodicity, suggesting that the procollagen may aggregate to form some collagen fibrils even before secretion. The major portion of the collagen matrix is probably expelled as procollagen, which assembles itself extracellularly.

Either at the cell membrane or later, nonhelical terminal portions of the procollagen are cleaved by an enzyme called procollagen peptidase. These nonhelical ends, called telopeptides, are thought to aid in the extrusion or alignment of procollagen. The smaller molecule remaining after the cleavage has a molecular weight of approximately 300,000 and is called tropocollagen.

Radioactively labeled hexoses have been used to verify that the osteoblasts are responsible for the synthesis of the protein–carbohydrate components of the matrix. Within 10 minutes after injection into rats, the label was observed in autoradiographs over the Golgi regions of osteoblasts. Within 4 hours, the prebone region was labeled, and some time later, the radioactive precursor had accumulated in bone tissue.

The Osteocyte

Osteocytes are easily identified by two characteristic features: (1) they reside within chambers called lacunae and (2) they exhibit numerous cytoplasmic projections radiating about in fine channels called caniculi. The extremely large area of the cell membrane–bone tissue interface obtained by this system is crucial to the control of calcium homeostasis, the principle function of osteocytes.

Many osteocytes undergo an identifiable life cycle that can span many years, beginning with envelopment and ending with cell death. Soon after the osteocyte is entrapped within its lacunae, the organelles still resemble and function like those in the osteoblast. Accordingly, this has been called the formative phase. As the cell matures, the Golgi complex and its associated vesicles become more prominent, suggesting a rise in lysosomes to aid in the resorptive phenomenon.

In a process unique to osteocytes, called osteolysis, mineral is withdrawn from and returned to the matrix surrounding the lacunae and caniculi, providing calcium for the minute-to-minute control of mineral homeostasis. During the process, the matrix is demineralized and often disoriented to the point of becoming flocculent when viewed by the electron microscopy. Should the stimulus for osteolysis be prolonged, several lacunae can become confluent, with the result that their osteocytes are freed.

The last stage of the osteocyte's life is degenerative, characterized by vacuolization of cell organelles, cell death, and disintegration.

The Osteoclast

The cell most responsible for the resorptive aspect of bone remodeling is the osteoclast. It is classically associated with erosion pits called Howship's lacunae resulting from a resorptive process called osteoclasis. Actually, the osteoclast is capable of removing bone from any surface with or without pit formation.

The osteoclast is the most fascinating of the bone cells, possessing not only an exceptional size and unusual presentation of organelles, but an enigmatic life history.

This bone cell can have multiple nuclei, ranging in number from two to over one hundred. The size of the cell increases proportionally, and as a result, the giant osteoclast can be the largest cell of the body. Besides the abundance of nuclei, the osteoclast is characterized by a plethora of mitochondria and a unique aggregation of centrioles called the centrosphere.

One of the most distinguishing features of an osteoclast is the specialized cell membrane found at the site of bone resorption called the brush border. It is a complex zone of folds and cytoplasmic projections contiguous to the resorbing matrix. The extensions are only fractions of a micron wide but may be several microns long. The spaces between these folds are tortuous and often show matrix debris between them when examined by electron microscopy. The folds terminate as vesicles or vacuoles that have been reported to contain apatite crystal. The process of bone resorption at the brush border probably entails physical as well as an enzymatic activities.

The origin of the osteoclast is controversial; its fate simply unknown. It is accepted that osteoclasts arise from the fusion of smaller cells, as no mitotic figures are ever seen. However, the source of these cells is unresolved. Some investigators maintain that the precursor cells of the osteoclast have the same lineage as that of osteoblasts, i.e., mesenchymal cells. Other evidence suggests strongly that osteoclasts derive from a fusion of wandering mononuclear phagocytes. The problem awaits future clarification.

The stimulus for osteoclasis is possibly related to electrical activity engendered in the bone by the bending associated with function. It is known that bone deformation leads to the development of electric potentials, presumably a piezoelectric effect. How such signals are translated into the many activities of osteoclasis has not been determined.

METHODS OF BONE FORMATION

Introduction

The deposition and calcification of bone tissue can occur in one of two ways, i.e., directly (intramembranous) or indirectly (cartilage replacement). In intramembranous formation, osteoblasts secrete a matrix composed of collagen and proteoglycans and additionally promote the deposition of calcium hydroxyapatite crystals into this matrix. The indirect method is more circuitous, utilizing a scaffold or precursor of cartilage on which to deposit the bone. In this situation, cartilage is first formed by chondroblasts, undergoes a degenerative process associated with mineralization, and then is invaded by bone resorbing cells, which reduce the cartilage to a framework. The osteoblasts in turn deposit bone matrix around this cartilage model. Eventually, the remnants of the cartilage matrix are completely lost in the process of growth and remodeling.

While the cartilage replacement mechanism is much less efficient than direct formation, nature has selected it for at least two important reasons. The first results from the limits on new bone formation imposed by the nature of bone itself. Because bone is calcified and hard, there is no possibility of internal expansion due to cellular proliferation or matrix deposition. This kind of growth, interstitial expansion, is characteristic of soft tissues, including cartilage. Bone as a tissue can only grow by apposition, a surface-limited process. Consider the handicap of such a restriction on the creation of a 7-foot basketball player from an 20-mm embryo. Cartilage, in contrast, does not calcify (until such time as it is to be converted to bone) and consequently has the potential for interstitial expansion. Thus, chondrocytes can undergo cell division and secrete matrix, rapidly forming new tissue mass, which is subsequently converted to bone (Fig. 2.2). An added advantage is the directionality of growth, which can be genetically programmed into some of the cartilage replacement mechanisms, such as the growth plates of long bones (to be discussed).

A second factor responsible for the indirect method of bone formation is the inability of bone cells to form new tissue while under pressure. This last statement does not imply that bone cannot withstand a load, which of course it can (we are not shapeless piles of protoplasm), nor does it mean that bone apposition cannot occur in a load-bearing situation, i.e., the thickness of the cortex of a long bone increases with age through the process of apposition. What it does say is that direct pressure on the periosteum, as might occur if the articular surfaces of two bones at a joint were composed of this tissue, prevents new bone formation and leads eventually to degenera-

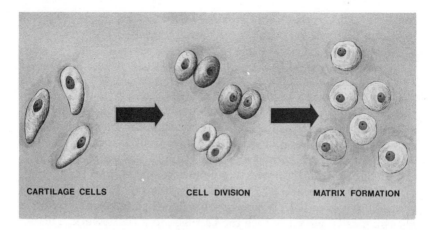

FIGURE 2.2 Interstitial expansion in cartilage resulting from cell division and new matrix formation.

tion and/or resorption. Cartilage is capable of bearing a load, and for this reason composes the articular surfaces of bone (Fig. 2.3). When this covering is damaged, as in arthritis, a crippling and painful condition results. Cartilage is capable of growth under load, because the cells involved are not on the surface as in bone but are buried within the tissue. Thus, they can undergo mitosis and secrete matrix removed from direct friction. In addition to this property of interstitial expansion, cartilage is not vascularized but is perfused by tissue fluids. Loads, then, cannot as easily obstruct the flow of body fluid to the chondrocytes.

The ability of cartilage to grow against a load is aptly demonstrated by the growth plates of the long bones of the legs, which continue to lengthen despite the weight imposed upon them by the torso and head.

Intramembranous Bone Formation

Intramembranous bone formation occurs on the outer surface of bone (the periosteum), the inner surface of bone (the endosteum), and in the case of a few bones in the skull, at the edges in specialized structures called *sutures*.

Periosteum and Endosteum

The periosteum is the tough fibrous tissue covering the outer surface of bone populated by osteoblasts and osteoblast precursor cells. For some period of time, the highly specialized inner zone of cells can

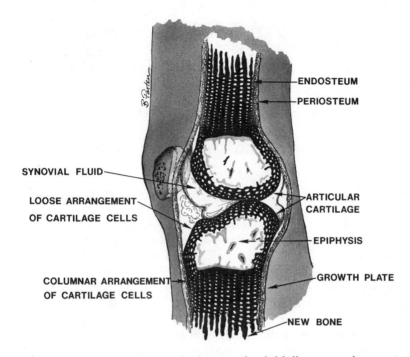

ENDOSTEUM
PERIOSTEUM
SYNOVIAL FLUID
LOOSE ARRANGEMENT
OF CARTILAGE CELLS
ARTICULAR
CARTILAGE
EPIPHYSIS
COLUMNAR ARRANGEMENT
OF CARTILAGE CELLS
GROWTH PLATE
NEW BONE

FIGURE 2.3   Cross section of a limb joint of a child illustrating the remaining areas of cartilage.

secrete matrix, and the periosteum drifts away as the bone thickens.
At some point, these osteoblasts enclose themselves in their own se-
cretions and become osteocytes. Other precursor cells differentiate
into osteoblasts to replace them and the process continues. When
resorption of bone is required during remodeling, periosteal osteoclasts develop from some primitive cell type.

The endosteum does not exhibit the capsular connective tissue
of the periosteum, but rather consists of a looser arrangement of osteoblasts and various pluripotential cells capable of differentiating
into bone, hematogenous, or connective tissue cell lines. The endosteum, then, is capable of apposition and resorption, and both do
occur during growth and development.

Sutures

The junctions between bones of the head are called sutures, and
these structures consist of soft tissue interposed between bones. The
soft tissue is composed of connective tissue and a variety of cell
types, including osteoblasts and osteoclasts. The most current con-

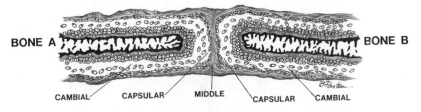

FIGURE 2.4   Diagram of a suture as a five-layered structure. After J.J. Pritchard, J.H. Scott, and F.G. Girgis. J Anat 90:73, 1956.

cept of the suture depicts it as a five-layered structure as illustrated in Fig. 2.4.

The cambial cells are precursor cells that replace the osteoblasts as bone deposition occurs. The purported existence of the capsular layers isolating a middle zone of neutral cells means that the synthetic process of two bones are isolated from one another. Bone A has its own battery of cells responsible for sutural deposition, and activity there is not necessarily reflected in similar cells of bone B. The division of labor in the suture allows for differential growth, whereby one bone of the skull might grow at a greater rate than its neighbor, because the brain expansion beneath them was not uniform. If it were not for this property of differential growth, our heads might resemble basketballs.

The older view of the suture, i.e., the middle zone cells proliferating and supplying osteoblasts to each bone and simultaneously creating an actual separating force, is not compatible with modern concepts of craniofacial growth.

Cartilage Replacement

Endochondral Bone Formation—The Growth Plate

There are three cartilage replacement mechanisms in the head, each of them without parallel in the rest of the body. But before discussing them, it might be wise to first present the process of endochondral bone formation, which is exhibited in every growth plate of every long bone of the body.

Each future limb bone begins embryonically as a cartilage anlage, or precursor (Fig. 2.5A). The cartilage cells are haphazardly arranged in this anlage, manifesting no specific orientation or direction of growth. Durkin has given the name *embryonic* to this type of cartilage.

Before long, the connective tissue around the cartilage, the periochondrium, thickens at the middle of the shaft forming a peri-

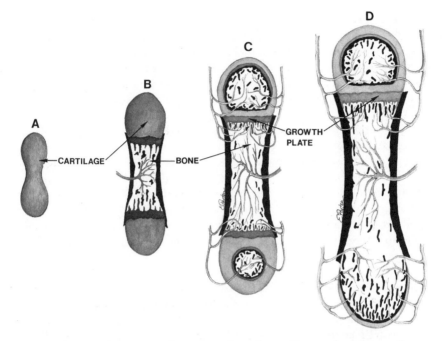

FIGURE 2.5   The steps in the conversion of a cartilage anlage to a mature long bone. (A) The cartilage anlage. (B) After invasion by the blood supply (periosteal bud) at the middle of the anlage (diaphysis), the calcifying cartilage is progressively converted into bone until only the ends remain cartilaginous (the epiphyses). (C) Some time later, the epiphyses are invaded by separate blood supplies and the ends become bone tissue except for the growth plates and articular surfaces (D, *upper*). When longitudinal growth ceases sometime after puberty, the growth plates are replaced by bone and the diaphysis and epiphyses become continuous (D, *lower*). Adapted by permission from P. Rubin: Dynamic Classification of Bone Dysplasias. Chicago, Year Book Medical Publishers, Copyright © 1964.

chondral ring. This new constriction diminishes perfusion, elicits mineralization, and eventually invites the invasion of a vascular bud (Fig. 2.5B). The pluripotential cells of the new blood supply differentiate into osteoblasts and osteoclasts, which begin to convert the cartilage to bone. The conversion spreads rapidly from the middle toward both the proximal and distal ends of the bone, but the process stops short, leaving the heads as cartilage. Somewhat later, these heads are invaded by separate blood supplies, and bone modeling follows (Fig. 2.5C). As a result, a section of cartilage is left sandwiched between bony compartments at each end of the bone; these remnants are responsible for all future linear growth and consequently are called growth plates (Fig. 2.5D). The very ends of the bones remain as cartilage, the tissue best suited for articulation.

The cells of the growth plate change from a random pattern into rigidly organized columns capable of exacting highly directional growth. According to Durkin, the embryonic character of the anlage cartilage has been transformed to a type now called specialized cartilage.

Briefly outlined, longitudinally aligned cells of the proliferative layer synchronously divide and secrete matrix (Fig. 2.6). As a consequence the cartilage growth plate expands interstitially, lifting the head of the bone away from the shaft and lengthening the whole bone. Because excessive cartilage lacks stiffness, some of the older cartilage must be converted to bone. This occurs on the shaft side of the growth plate in an elaborate process involving degeneration of chondrocytes, calcification of the cartilage, erosion by bone cells, and eventually bone deposition.

This active separation of tissues against a load is characteristic of endochondral growth. The growth plate and other bone-forming areas with similar properties are called growth centers, a term introduced by Baume. Diametrically opposed to the autonomous nature of growth centers is the adaptive, passive nature of other bone-forming areas. In these situations, new bone is formed only as a filling-in process when the separation of the bones is instigated by other forces. Such locations are known as growth sites, and the suture is almost universally accepted as being one.

Just as gray lies between black and white, there are bone-forming units that demonstrate properties of both growth centers and growth sites, and the labeling of a number of these located in the head is presently very controversial.

Spheno-occipital Synchondrosis

The structure on the skull most resembling the growth plates of long bones is the cartilage remnant located between the sphenoid bone and the occipital bone along the midline of the cranial base. Its cellular organization suggests the appearance that one might get by butting together the reserve cartilage layers of two growth plates (Fig. 2.7). As a result of this architecture, interstitial expansion is bi-directional, increasing the size of bones simultaneously.

Because the spheno-occipital synchondrosis looks like two growth plates back to back, it has been tempting to simply call it a growth center capable of independent tissue separation. However, experiments comparing the growth of both transplanted synchondrosis and growth plates have indicated that the growth plates exhibit more autonomous growth. This suggests that part of the

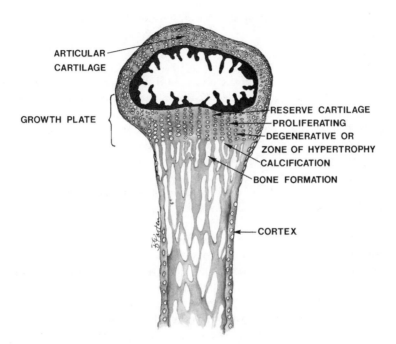

FIGURE 2.6   The zones of cells in the growth plate that are responsible for the conversion of cartilage to bone.

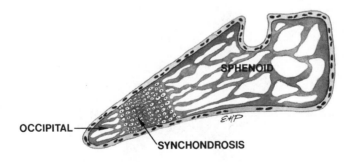

FIGURE 2.7   The spheno-occipital synchondrosis situated between the sphenoid and occipital bones at the midline of the cranial floor.

impetus of complete synchondrotic growth may arise from brain growth. Unfortunately, the question concerning the autonomous or adaptive nature of the spheno-occipital synchondrosis has not been resolved.

Nasal Cartilage

The nasal septum seen in the adult skull divides the nasal cavity into two parts (Fig. 2.8). It runs vertically in the midline of the face from the ethmoid and sphenoid above to the vomer below. Initially, this septum is formed entirely of cartilage, but in the adult, all of it is converted into bone save the anterior segment uniting with the tissues of the nostril. The advancing bone tissue sweeps diagonally across the cartilage, downward and forward.

The cartilage of the nasal septum is presumably capable of interstitial expansion, and because the vertical size of the nasal cavity increases so dramatically during growth, much speculation has naturally arisen about the role of the nasal cartilage in the process. Both generalized division of chondrocytes and/or regional mitosis along the zone of conversion could create a tissue-separating force that would act to propel the maxilla and contiguous bones downward and forward. If this force were in fact generated, the nasal septum could be classified as a growth center. Unfortunately, experiments designed to test the role of the nasal cartilage have been equivocal, and its clearcut designation as a growth center, growth site, or something in between, has not been possible.

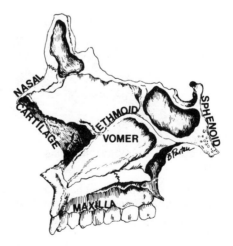

FIGURE 2.8 A midsagittal section showing the anatomic relationship of the nasal cartilage to the cranial base and the maxilla.

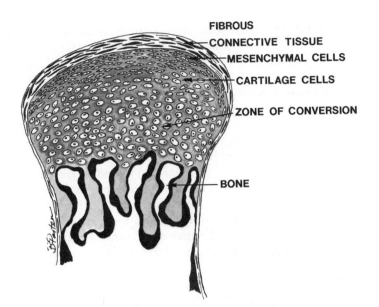

FIGURE 2.9 The head of the condyle showing the unique arrangement of tissues and cells.

Condyle

In the long bone, the load-bearing surface, the articular cartilage, is separated from the cartilage growth zone by epiphyseal bone. In the condyle, on the other hand, the articular surface and the cartilage replacement tissue are juxtaposed. In addition, their histologic character and source of cellular replenishment are unlike those found at any other site.

Most of the articular surfaces of long bones are cartilage that can expand internally as a result of cell divisions in more centrally located zones. The growth in thickness of the articular surface compensates for the loss of tissue resulting from functional wear and tear. The most peripheral zone of the condylar articulation is not composed of cartilage but rather of connective tissue (Fig. 2.9).

Immediately subjacent to this outer layer is a zone of mesenchymal cells capable of two functions. First, they can secrete the constituents of the overlying fibrous layer, and second, they can differentiate into cartilage cells to replenish those that are sacrificed in the conversion to bone, a process that occurs in the opposite direction.

In the condyle, a replacement mechanism is operating during growth that differs from other areas both in the source of new cartilage cells and in the orientation they assume. In these units, new cartilage cells are derived from noncartilage precursors; in all other replacement mechanisms, new cartilage cells are derived from other cartilage cells by division. The cells of the cartilage of the condyle remain haphazard, random, or embryonic; they do not form the specialized columnar arrangement of the growth plate.

At least one authority has speculated that the unique character of the condyle is not so much a genetically enforced structure as it is an adaption of the periosteum to load. Such a theory implies that periosteal pluriopotential cells can revert to chondrocytes in vivo as they have been shown to do in tissue culture experiments. Pressure on the articular surface of the condyle would dedifferentiate the periosteal cells and effect a semifibrous and semicartilage covering. Whether such a concept is valid remains to be answered, but without question, the condyle is unique as a cartilage replacement mechanism.

MECHANISMS OF BONE TRANSFORMATION

The previous section dealt with those mechanisms that were concerned with the development of form and size—the morphology of the bone. This process is limited by maturation; during puberty or some years thereafter, our maximum size is attained and our bones no longer grow. They do not cease to change, however. The development of their internal architecture or structure continues throughout life in response to the change in stress upon them. For instance, if we were to put on a middle-age spread, our bones would adapt to the added load.

This transformation also occurs during bone formation, because, as will be seen, remodeling (a type of transformation) is an integral part of bone growth. Superimposed on this process are the changes in stress associated with growth.

Compact Bone

Two compartments are operative in bone transformation: the compact bone and the spongy bone (Fig. 2.10). Compact bone, or cortical bone, is sandwiched between the periosteum and endosteum and can increase in size or drift in some direction by appropriate apposition and resorption. These latter activities are responsible for external transformation. Internal transformation of compact bone involves the development of haversian systems, which consist of multiple cylinders of new bone running in suitable directions throughout the original bone (Fig. 2.11). These cylinders, called

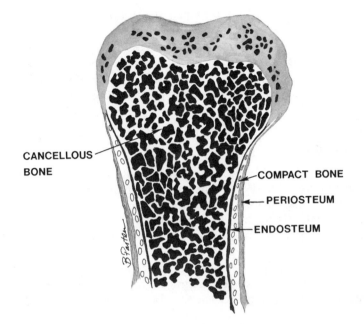

FIGURE 2.10   The two principal morphologic divisions of bone—compact and cancellous.

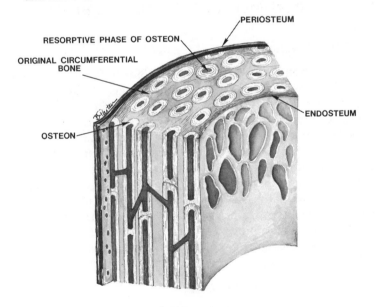

HAVERSIAN SYSTEM

FIGURE 2.11   The haversian system of mature, compact bone.

osteons, are formed by an internal erosion followed by a redeposition. What remains are cores of new bone up to 250 μm in diameter and 5000 μm in length that house blood vessels in their centers. This "secondary" vascular system serves two purposes: first, an adequate blood supply is insured to all the osteocytes of bone, and second, new stresses can be handled by the proper orientation of these new bone formations.

Spongy Bone

The insides of most bones have porous texture described as spongy or cancellous. This tissue consists of interconnecting trabeculae of bone interspersed with various kinds of bone cells, connective tissue, hematopoietic tissue, and fat. Spongy bone, particularly around the head of a long bone, has a internal orientation that seems a perfect adaption to the stresses placed on that bone.

The dividing line between compact and spongy bone is by no means immutable; through the process of growth and development compact bone has many occasions to become spongy, and vice versa.

EMBRYOLOGICAL ORIGIN OF THE CARTILAGE REPLACEMENT MECHANISMS OF THE HEAD

The Chondrocranium

Most of the bones associated with the floor of the cranium and the cranial base begin their existence as cartilage. During embryologic development, areas around the notochord begin to chondrify, and eventually most of the occipital, the petromastoid portions of the temporals, the majority of the sphenoid, and a great deal of the nasal area including the ethomoid are aggregated into a cartilaginous complex called the chondrocranium. At birth, very little of this primordial cartilage remains, having been converted into bone. The junctions between these cartilaginous structures, and between them and membrane bones, are for the most part demarcated by sutures. Two prominent exceptions to the complete postnatal elimination of cartilage are the spheno-occipital synchondrosis and the nasal cartilage.

Spheno-occipital Synchondrosis

The sphenoid bone consists of a central body and lateral greater and lesser wings. The body is divided embryologically into a presphe-

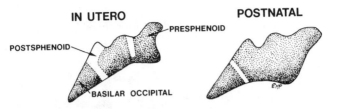

FIGURE 2.12   The formation of the spheno-occipital synchondrosis following ossification of the original chondrocranium.

noid part (that which is anterior to the tuberculum sellae) and the postsphenoid segment, comprising the sella turcica and dorsum sellae (Fig. 2.12). These two parts fuse by 8 months of fetal life. Posterior to the postsphenoid, the cartilage of the basilar part of the occipital bone is simultaneously becoming ossified. Both the postsphenoid and basilar occipital continue ossification until all that remains is a plate of cartilage between them, the spheno-occipital synchondrosis. This area is responsible for most of the lengthening between the foramen magnum and sella turcica, a process that continues until fusion of the sphenoid and occipital bones during the latter half of the second decade of life.

Nasal Cartilage

The embryologic nasal capsule develops into the ethmoid bone and the inferior nasal conchae. The ethmoid bone consists of a horizontal or cribriform plate, which forms a part of the cranial base; a perpendicular plate, constituting part of the nasal septum; and two lateral masses of labyrinths (Fig. 2.12). The perpendicular plate and the christa galli of the cribriform begin to ossify, forming a single center during the first year after birth. Ossification continues until they join the already ossified labyrinths sometime during the second year. Thus, at this time, a septum with potential tissue-separating force is buttressed posteriorly against the sphenoid and superiorly against the cribiform plate, just opposite to an anterior and inferior articulation with the vomer. This arrangement makes it easy to visualize a downward and forward thrust of the maxillary complex as a result of nasal cartilage expansion.

Splanchnocranium–Condylar Cartilage

The maxilla and mandible, intramembranous bones, are derived from the first branchial arch. The maxilla has no cartilage precursor whatsoever, but the mandible develops membrane bone lateral to a

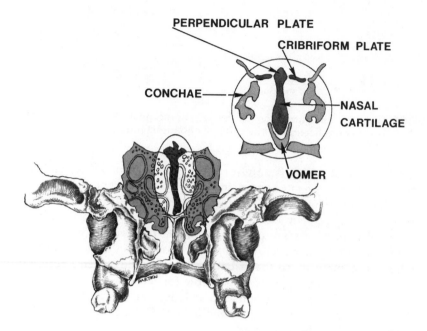

PERPENDICULAR PLATE

CRIBRIFORM PLATE

CONCHAE

NASAL CARTILAGE

VOMER

FIGURE 2.13   Cross section of the midface showing the relationship of the nasal cartilage to the floor of the cranium (*top*) and the vomer and maxilla (*bottom*). Adapted from J.H. Scott: Dento-facial Development and Growth. Oxford, Pergamon Press, 1967.

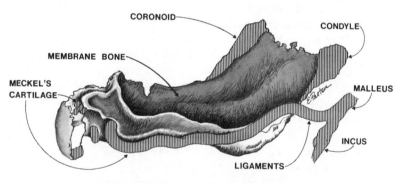

CORONOID

CONDYLE

MEMBRANE BONE

MECKEL'S CARTILAGE

MALLEUS

INCUS

LIGAMENTS

DEVELOPMENT OF THE MANDIBLE

FIGURE 2.14   The relationship of Meckel's cartilage to the membrane bone and the secondary cartilage of the coronoid and condylar processes.

temporary structure called Meckel's cartilage (Fig. 2.14). This cartilage contributes to the formation of the two auditory ossicles (malleus and incus), to the sphenomandibular ligament, to the spine of the sphenoid, and to a small segment of the chin. Meckel's cartilage is completely resorbed from the mental foramen to the lingula, and as a result makes only a small contribution to the mandible proper. The cartilage that is found in the condyle is derived from mesenchymal cells unrelated to the first branchial arch (Meckel's) or to the chondrocranium. It is often referred to as a secondary cartilage, since its formation is secondary to the original primordial cartilage. Durkin suggests that the term secondary has little descriptive value and should be replaced with embryonic, since the organization of the cartilage cells of the condyle resembles that of the anlages of long bones and not the specialized growth plates.

Skeletogenesis and Facial Development

Macromolecular Transition in Extracellular Matrix

Although very little information has accumulated on the biochemistry of human facial cartilage and bone, recent investigations of limb formation in the chick provide insight into the fundamental role of extracellular matrix in skeletal morphogenesis. Similar activities in human facial development very likely occur.

Zwilling divides the formation of the cartilage anlage of the limb and its subsequent conversion to bone into two phases: morphogenesis and cytodifferentiation. The former is characterized by proliferation and interactions that determine the future shape of the limb, the latter by differentiation and specialization of cells and tissue.

During the morphogenetic phase, the limb bud is characterized by a distinct extracellular matrix. The primitive mesenchymal cells elaborate type I collagen and abundant hyaluronate. Meager amounts of low molecular weight chrondroitin sulfate proteoglycans are synthesized.

As "condensation" of the core begins, the enzyme hyaluronidase, which is capable of digesting hyaluronate, dramatically increases in concentration. Coincident with the rise of enzyme activity, the tissue content of hyaluronate falls rapidly. In its stead appears a type II collagen and a large molecular weight chondroitin sulfate proteoglycan.

Condensation ushers in the phase of cytodifferentiation and the appearance of the temporary rudiment of the limb, the cartilage anlage. In the conversion of this latter structure to bone, the cells revert again to the synthesis of a type I collagen. In addition, much of the proteoglycan is lost during the calcification process.

Role of the Extracellular Matrix in Skeletogenesis

It would seem inefficient for nature to undergo major transitions in extracellular macromolecules for the sole purpose of providing structural support. We can safely assume that such turnover is also associated with the control of morphogenesis. Several diverse findings do, in fact, support the theory that matrix helps control cell migration, recognition, mitosis, and differentiation.

The phenotypic expression of chondrocytes in vitro seems to be promoted more by one substrate than another. For instance, when somites are cultured either on a type I or type II collagenous substrate, the production of proteoglycans and new collagen by the somite chondrocytes is enhanced by type II collagen.

Proteoglycans have also been shown to possess stimulatory activity in cultures of somites and notochord. The induction of somite chondrogenesis is thwarted by enzymatic removal of the proteoglycan coat of the notochord, whereas somites alone can be stimulated to produce cartilage by the addition of cartilage proteoglycans.

Thus, it appears that the major extracellular macromolecules, namely, type II collagen and chondroitin sulfate proteoglycans, while not specifically effecting the differentiation of chondrocytes, cooperate to stabilize them and promote their synthetic activity.

The prominence of hyaluronate in the chick embryo limb and its rapid decline during cytodifferentiation suggest that it may play an important role early in morphogenesis. Because hyaluronate has been shown to undergo high degrees of hydration during the embryologic development of several organ systems, this property has been linked to morphogenesis. As mentioned previously, the negative charge of the glucuronate moiety could create an environment conducive to the exertion of an osmotic pressure, a phenomenon leading inevitably to swelling. A tissue in which this occurred would be open to the migration of mesenchymal cells. When the cell population was sufficiently dense, hyaluronidase secretion would begin the breakdown of hyaluronate. As the matrix collapsed around the cells, the environment would change and the process of cytodifferentiation could begin.

A Working Hypothesis of Facial Skeletogenesis

From the kinds of findings just discussed, Toole and Linsemeyer have developed a working hypothesis of limb skeletogenesis; although there is no direct evidence of like activities in the human

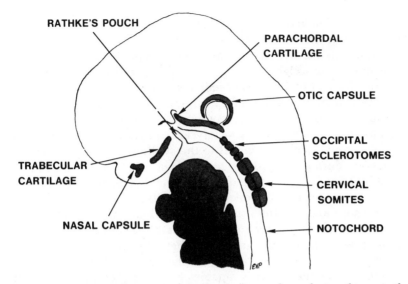

FIGURE 2.15 Sagittal section of the head of an embryo during the seventh
week of development. Adapted from G.H. Sperber: Cranio-facial
Embryology. Baltimore, Williams and Wilkins, 1973.

chondrocranium, it is not unreasonable to speculate about such involvement in facial morphogenesis.

The primordial cartilages of the chondrocranium begin to form
toward the end of the embryonic period, somewhat later than the
primordia of the brain, the cranial nerves, and the eyes. The initial
step involves the concentration of occipital sclerotomal mesenchyme
around the notochord, which underlies the developing hindbrain
(Fig. 2.15). As the mesenchymal concentration extends cephalically,
a floor for the brain is formed. Conversion of this condensation into
cartilage signals the beginning of the chondocranium. Chondrifica-
tion centers around the notochord are called parachordal cartilages,
and their caudal extension incorporates the fused sclerotomes of the
occipital somites, forming the anlagen for the basilar and condylar
parts of the occipital bone.

The hypothesis for craniofacial development includes the following sequence of events. The sclerotomal mesenchymal cells are
stimulated to migrate around the notochord in response to some inductive or chemotactic factor. These cells synthesize type I collagen,
a small molecular weight proteoglycan, and abundant hyaluronate.
The hyaluronate-rich matrix swells, providing an attractive milieu
for the migration and proliferation of cells. Precocious differentiation is prevented by type I collagen.

The sudden loss of hyaluronate associated with the elevation of hyaluronidase activity promotes the differentiation of mesenchymal cells into cartilage cells. The changeover to type II collagen and new large molecular weight proteoglycans stabilizes the differentiation process and acts as a positive feedback to further matrix synthesis. Some time later, when the chondrocranium begins to ossify, further transitions of collagen and proteoglycans occur. At the site of endochondral osteogenesis, the proteoglycans are suddenly depleted and the cells, now osteoblasts, revert to the synthesis of type I collagen. Somehow, during the process of cartilage calcification and subsequent invasion by vascular and mesenchymal cells, inductive factors promote a modulation to bone-forming cells.

In those areas of the face and skull where membrane bone is formed, we can at this time only speculate that the position of the presumptive bone tissue, sandwiched between the matrices of skin and muscle, eliminates the cartilage replacement step. We know that demineralized bone matrix is capable of inducing osteogenesis when transplanted either intramuscularly or subcutaneously. This property seems contingent on a protein associated with, or possibly covalently linked to, collagen. We can then surmise that the special environment of these regions prompts the mesenchymal cells to synthesize a specific protein linked in some manner to collagen. Even a subtle variation as this to the extracellular matrix could promote further osteogenesis instead of chondrogenesis, causing bone to form directly.

Experimental results have suggested that induction by brain tissue plays a role in the osteogenesis of cranial bone. When portions of chick embryo brain were removed early in development, the corresponding bones that normally covered that segment of brain were reduced or missing. When different regions of chick brain tissue were grafted to the chorioallantoic sac of the developing chick embryo, the corresponding cranial bones formed. These studies offer convincing evidence that the brain is capable of inducing the calvarium. They also support one aspect of the functional matrix theory (see Chapter 6), which contends that there are no cells genetically predetermined to become bone cells. Rather, pluripotential cells are so induced by extragenic influences, electrical and/or chemical.

BIBLIOGRAPHY

Arey LB: Developmental Anatomy, 6th ed. Philadelphia, Saunders, 1954
Babula WJ, Smiley GR, Dixon AD: The role of cartilaginous nasal septum in midfacial growth. Am J Orthod 58:250, 1970
Baume LJ: Principles of cephalofacial development revealed by experimental biology. Am J Orthod 47:881, 1961

Durkin JF: Secondary cartilage: a misnomer? Am J Orthod 62:15, 1972

Enlow DH: Handbook of Facial Growth. Philadelphia, Saunders, 1975

Ford EH: Growth of the human cranial base. Am J Orthod 44:498, 1958

Grant ME, Prockop DJ: The biosynthesis of collagen. N Engl J Med 286:194, 1972

Ham AW: Histology, 7th ed. Philadelphia, Lippincott, 1972

Koski K: Cranial growth centers: facts or fallacies? Am J Orthod 54:566, 1968

Lacroix P: The Organization of Bones. London, Churchill, 1951

Pritchard JJ, Scott JH, Girgis FG: The structure and development of cranial and facial sutures. J Anat 90:73, 1956

Ranly DM: Bone apposition and resorption. In Lazzari EP (ed): Dental Biochemistry, 2nd ed. Philadelphia, Lea and Febiger, 1976

Rubin P: Dynamic classification of bone dysplasias. Chicago, Year Book Medical, 1964

Scott JH: The cartilage of nasal septum. Br Dent J 95:37, 1953

Sperber GH: Craniofacial Embryology. Baltimore, Williams and Wilkins, 1973

Storey E: Growth and remodeling of bone and bones. Am J Orthod 62:142, 1972

Toole BP, Linsenmayer TF: Newer knowledge of skeletogenesis. Clin Orthop 129:258 1977

Urist M, Iwata H, Strates B: Bone morphogenetic protein and proteinase in the guinea pig. Clin Orthop 85:275, 1972

Zwilling E: Morphogenetic phases in development. Dev Biol suppl 2:184, 1968

· 3 ·

Principles of
Bone Growth

One does not arrive at fundamental laws and principles as a function of what is already known. Such laws and principles do not merely explain; they illuminate. They do not merely add to what we know; they create a new space in which knowing can occur.

Werner Erhard

INTRODUCTION

A femur is a femur and a humerus is a humerus, and a forensic pathologist or anatomist would be able to differentiate between them. Although these two bones have enough dissimilarities to allow for such discrimination, they are basically variations on the same theme. Both are linear, with shafts and two heads; both have articular surfaces; both have cortices and spongy bone; and both increase in size by growth plates and remodeling.

If they have so much in common, is it safe to assume that both bones grow by very similar processes? Probably. It is unlikely that nature would elaborate different methods of bone formation for every bone; it would be more economical to modify a basic process when needed. If it is true that a fundamental process exists, then it is probably also true that certain laws of nature govern this system.

The understanding of these laws is one of the coveted objectives of the bone biologist.

We are nowhere near understanding the real laws that control the size and shapes of bones. We have some knowledge of collagen formation by osteoblasts; we understand some facets of mineralization; we are aware of the influence of soft tissue on bone; we speculate on electrical signals being able to generate chemical responses in cells. But in fact, we know very little about why a bone is and why it looks as it does.

What is evolving are working concepts to deal with these questions. The rest of this chapter is concerned with descriptive principles of bone growth. These principles allow us to reduce the description of the growth of many variously shaped bones to a few common denominators. These principles or assumptions are formulated to help us; nature very likely operates from an entirely different set of laws. Only time and much research will reveal the true laws, but if these tentative principles assist in describing and understanding growth in the meantime, then they are worthy of our analysis.

We are indebted to Donald Enlow for the precise expositions of two rather old principles of bone growth, i.e., drift versus displacement and posterior growth–anterior displacement, and for the synthesis of a new principle describing the relationship between the surface of bone deposition and the direction of bone growth.

DRIFT VERSUS DISPLACEMENT

In modern cephalometrics (the science dealing with the measurement of the roentgenographic image of the head) many anatomic landmarks are selected for analysis. Let us measure for our purposes the distance between the center of the sella turcica and the point of the chin (gnathion). If we were to compare this distance found in cephalograms of the same individual, for example, at the age of six and again at ten, we would invariably find that, excluding pathology, this distance would increase with age. Gnathion would move away from sella turcica, but the measurement would only tell us how much or what the rate was. It would not tell us if this change occurred because osteoblasts deposited new bone on the chin or whether the condyle grew and pushed the chin forward. In fact, both phenomena participate. The deposition process is called drift; the "push" contribution is called displacement.

Drift is the movement of bone in space by virtue of apposition on one side of a cortex or surface and resorption on the other. It can be likened to the movement of a sand dune where grains of sand are blown by the wind from the one side to the other; the sand dune is drifting. If you drove up in your steam shovel, picked up the sand dune, and moved it to another location, then you would have displaced it.

When gnathion (Gn) moved away from the cranial base, it did so by apposition directly on the chin point (drift) and because growth at the condyle relocated it forward (displacement). Most growth in the head is a combination of these two processes; there are few examples of either pure drift or pure displacement.

POSTERIOR GROWTH-ANTERIOR DISPLACEMENT

Let us imagine that you are floating on your stomach at the edge of a swimming pool with your legs folded. Wishing to propel yourself to the middle of the pool, you might place your feet against the side and rapidly straighten your legs (Fig. 3.1). Suddenly, your body would thrust forward, and you would have demonstrated a "posterior growth–anterior displacement." Your legs "grew" in a sense, and because your feet were lodged against an immovable wall, your body was pushed in the opposite direction. Your feet moved away from the body toward the wall, and this movement was expressed by

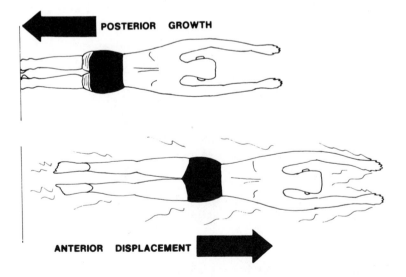

FIGURE 3.1 A swimmer propelling himself from the side of a pool—an example of posterior growth–anterior displacement.

head movement away from the wall. In anatomic terms, the feet moved posteriorly and the head anteriorly.

Let us return to the mandible. We have already described that the chin point moves away from the sella turcica by a combination of drift and displacement. Of the total increase, the contribution of anterior apposition is minimal. By far, most of the lengthening process of the mandible occurs on the ramus and at the condyle, and since the condyle articulates against an immovable wall (the glenoid fossa), the mandible is displaced anteriorly.

This principle of backward growth–forward movement is manifested in every long bone of the body and in a number of bones of the head.

DIRECTION OF BONE GROWTH AND THE CHARACTER OF BONE ACTIVITY

Surface Apposition and Resorption—Its Relationship to the Direction of Bone Growth

This innovative concept of Enlow greatly facilitates the description of the apposition and resorption patterns that are observed in complex bones such as the mandible and maxilla. It also requires some

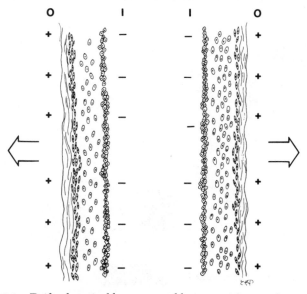

FIGURE 3.2   Drift of cortical bone caused by apposition on the outside (O) and resorption on the inside (I).

imagination and latitude in its application. In essence, the postulate states that the surface of new bone formation always faces in the direction of growth. Stated another way: If you can determine the direction of growth of any anatomic surface of a bone, you will automatically know that new bone formation is associated with that surface. Take as an example the simple situation of the increase in diameter of the shaft of a long bone during growth (Fig. 3.2). In Fig. 3.2, the cross-sectional thickness of the shaft is increasing; the direction of growth in this two-dimensional picture is lateral. In keeping with Enlow's principle, new bone formation should be associated with the outside areas (0). This is in fact where periosteal apposition occurs.

To be consistent with the foregoing concept, the surface of bone away from the direction of growth must undergo resorption. This phenomenon is invariably observed. The endosteum or inside (I) of the cortices are resorptive. The net result of periosteal apposition and endosteal resorption is the drift of the bone laterally. Thus, a bone increases its width by adding matrix to the surface facing the direction of growth and subtracting matrix from the surface away from the direction of growth. The two processes are not necessarily always equal; in the case of the shaft of long bones, apposition exceeds resorption, and the cortex thickens while drifting.

The V Principle

Unfortunately for our conceptualizing, the direction and the surface are not always so distinguishable. For example, the growth process of the head of a long bone shown in Fig. 3.3. As depicted here, the head of the bone is moving away from the shaft, and as a result, those shaded areas find themselves resorbed during the remodeling process. In this situation, the direction of growth and the surfaces undergoing apposition and resorption are not readily apparent. To simplify matters, Enlow has observed that many surfaces are better expressed as Vs than flat surfaces. Let us apply a little imagination to the study of the head of a long bone (Fig. 3.4). The direction of growth is, as the arrow indicates, away from the shaft. The surface of new bone deposition is, if the point is stretched, the growth plate that is represented by the inside of the V. In this situation, the inside of the V is where the new tissue is formed; it also represents the surface facing the direction of growth.

In barest terms, growth of the head of a long bone can be expressed as a series of ever-increasing Vs (Fig. 3.5), and by superimposing the preceding two figures, we obtain a visual perspective into

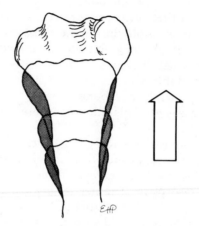

FIGURE 3.3   The growth process of a long bone showing the elongation in the direction of the arrow and the remodeling resorption involved (dark areas). Adapted from Enlow: The Human Face. New York, Harper and Row, 1968.

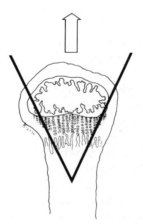

FIGURE 3.4   Visualizing the epiphyseal complex (growth plate) as a configuration resembling a V growing in the direction of the arrow.

the relationship between a complex surface and the direction of growth (Fig. 3.6). Using this concept, it is easy to see why the shaded areas are resorbed: they represent surfaces away from the direction of growth.

There are numerous structures in the head that can be visualized as a V, and when such an orientation is correlated to the direction of growth, immediate insight into patterns of apposition and resorption are gained.

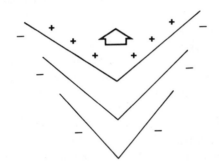

FIGURE 3.5 An ever-enlarging V brought about by addition on the inside (in the direction of growth) and subtraction on the outside (away from the direction of growth). Adapted from Enlow: The Human Face. New York, Harper and Row, 1968.

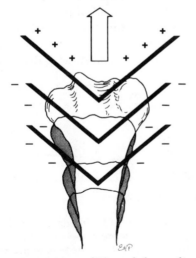

FIGURE 3.6 A superimposition of Vs and the outline of growing epiphysis, illustrating that new tissue formation can be imagined on the inside of the V (toward the direction of growth) and resorption on the outside (away from the direction of growth).

THE LAWS OF ELECTROGENESIS

Contradictions of Bone Growth

Throughout the body, one can find numerous apparent contradictions related to the effect of force and the reaction of bone to that force. Consider, for example, that added compressive forces on the heel bone (which might occur with a weight gain) results in added bone formation; compressive force against a tooth results in resorption of bone and tooth movement. Consider this contradiction: Bone is moldable under certain external forces, yet it reaches a normal configuration under physiologic loads and muscular pulls. As examples of bone adaptability, let us cite the following: malformation of the skulls of certain African tribes as a result of head bindings, the effect of leg braces in reshaping bones, and the dramatic alteration in the shape of the mandibles of patients wearing Milwaukee braces for the spinal condition scoliosis (this brace pits the occipital bone and the mandible against the pelvic bone to keep the spine straight).

It is not difficult to visualize how these extraneous forces can deform bone, but what might be difficult to understand is why physiologic muscle pull does not engender the same reaction. Contraction of muscle creates tension that is distributed to the periosteum via tendon attachments. Apparently, the stimulus for osteoblastic activity is not simply this tension acting directly, because often times new bone formation is in a direction opposite to muscle pull. As an example, the angle of the mandible becomes more acute with age despite the pull of the masseter in a direction that almost bisects that angle (Fig. 3.7A, left). Reasonably, the pull of

FIGURE 3.7 (*Left*) An illustration of the manner in which the gonial angle of the mandible becomes more acute during growth despite a constant muscle pull in the direction opposite to growth. (*Right*) An illustration of mandibular distortion (gonial angle becoming more obtuse) that might be expected to occur considering a simplistic relationship between bone and force. After Frost: The Laws of Bone Structure. Springfield, Illinois, Thomas, 1964.

this muscle should increase the angle between the body and the ramus (Fig. 3.7B, *right*). If bone is truly moldable (and there is ample evidence that it is), why, then, don't the forces generated by muscle contraction, which are as real as those of braces, cause bizzare patterns of formation?

To answer this question, Frost, like Enlow has formulated some laws that appear to operate within the world as we know it. We must reiterate, however, that the reasons that bone behaves as it does and the reasons that we conjure up are not necessarily the same. But as long as our contrivances remain consistent to the tests given them, they remain useful tools for further research.

The Relationship of Bending and Surface Curvature

Frost has developed what can be called the "laws of electrogenesis." This concept equates the change in the surface curvature of a bone under load to the kinds of cell reaction triggered. According to this theory, bone that is loaded normally is so structured that no bending occurs. In such a system, osteoblastic formation and osteoclastic reabsorption are in balance, and therefore, the bone does not change in response to the load. If loading occurs that is sufficient to bend the bone, then certain signals are generated (presumably electrical) that predispose the surfaces made more concave to osteoblastic activity and those surfaces made more convex to osteoclastic activity. Bending upsets the balance of formation (+) and resorption (−), and the bone cells attempt to remodel the bone in such a way that the new load no longer bends the bone and all cell activity returns to equilibrium.

Consider the reaction of a long bone during the process of weight gain (Fig. 3.8). The added load bends the bone slightly; the concave surface manifests an inhibition of osteoclastic activity, and osteoblastic formation predominates. On the surface that is made more convex, osteoclasis predominates over osteoblastic activity. The result of these surface-related phenomena is the straightening of the bone and an adaption to the increased load.

As an example of this law operating in the head, let us return to the situation of the mandible and the masseteric forces. To apply the principle, we must consider the angle of the mandible as a surface (Fig. 3.9). When the muscle contracts, this surface becomes less convex; less convex is a way of saying more concave. According to the law of electrogenesis, more concavity incites osteoblastic predominance. Thus, the angle of the mandible becomes more acute (depository) as a result of surface bending.

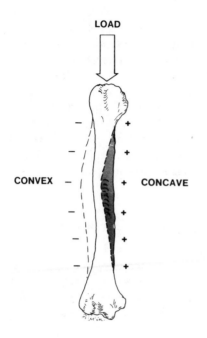

LOAD

CONVEX — | + CONCAVE

FIGURE 3.8   When sufficient load is applied to a bone, bending occurs. A differential pattern of electrical signals elicits osteoblastic activity on the surface made more concave and osteoclastic activity on the surface made more convex. After Frost: The Laws of Bone Structure. Springfield, Illinois, Thomas, 1974.

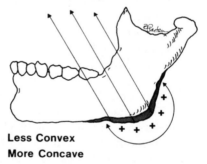

Muscle Contraction

Less Convex
More Concave

FIGURE 3.9   Based on the phenomenon of surface activity and bending, the direction of growth of the angle of the mandible opposite to that of muscle contraction can be explained. After Frost: The Laws of Bone Structure. Springfield, Illinois, Thomas, 1964.

BIBLIOGRAPHY

Enlow DH: The Human Face. New York, Harper and Row, 1968
Frost HM: The Laws of Bone Structure. Springfield, Illinois, Thomas, 1964

· 4 ·

Description of Growth by Anatomic Divisions

It is the common wonder of all men, how among so many millions of faces there should be none alike.

Sir Thomas Browne

AGE-RELATED CHANGES OF FACIAL APPEARANCE

Day by day, year by year, almost imperceptibly, the face of the newborn child is transformed by growth and by aging; inexorably, the cute, pug-nosed, bland face of infancy assumes in turn the expression of youth, the individuality of adulthood, and the character of age. The first 20 years are anabolic, growing and developing; the years after are maturing, catabolic and degenerative.

The infant enters the world with a brain and eyes more fully developed than the rest of his body, and therefore, his eyes dominate a small face tucked beneath a prominent cranium. The face appears broad and flat; the lower jaw seems underdeveloped and receded. The broadness results from the lack of vertical growth that is yet to come; the horizontal dimensions are more nearly adultlike. The floor of the nasal cavity is on the same level as the lower orbital rims, a position that changes dramatically in the ensuing years. The tooth-less arches and diminutive mandible contribute little to vertical height. The flatness is caused by a small, virtually bridgeless nose; upper orbital rims as yet unexpanded by frontal sinuses; and lateral orbital ridges and cheek bones that are still positioned forward with the nose. Lacking a definitive structure between them, the eyes appear wideset and prominent.

With age, the face undergoes "catch-up" vertical growth that changes its proportions with respect to the cranium. The middle face emerges from beneath the forehead and grows downward with great alacrity. Teeth erupt and the mandible swings forward and downward to maintain the occlusion, and the angles of the ramus increase laterally, "squaring up" the lower jaw. Such activities reduce the broadness and usher in the adult face.

The increase in the size of the nose, both soft tissue and bony bridge, combined with forward-growing supraorbital rims as a result of frontal sinus expansion, diminish the dominance of the eyes. Continued growth of the midline of the face and a lagging behind of the lateral orbital rims and zygomatic processes tends to create a convex shape. This loss of flatness is more characteristic of some individuals and some races than others. Mandibular growth and symphyseal apposition create a chin point, particularly in the male.

After adolescence, changes in the face can be ascribed mostly to soft tissue changes; only in cases of severe tooth loss are there pronounced alterations of bone. Edentulous arches lack alveolar processes and a vertical dimension that promotes normal muscular function. Loss of interocclusal distance upsets muscular balance and leads to bone resorption, mostly in the mandible.

With age, the fleshy nose tends to enlarge, the texture of the skin changes, wrinkles appear, and the soft tissue of the face can sag.

CRANIOFACIAL GROWTH RATES AND CHANGING PROPORTIONS

The differential pattern of craniofacial growth is graphically shown when the percent completion is plotted as a function of age (Fig. 4.1). The cranium, as represented by cranial width and length, is nearer its adult size than any other part of the head. This can be explained by the development of the brain, which by the eighth month of intra-uterine life has all the nerve cells it will ever have; myelinization accounts for the bulk of brain expansion after birth. A similar growth pattern of the eye, a brain-related sense organ, is evidenced by the early expansion of the orbit, as measured by orbital height. By 10 years of age, the calvarium and orbit have virtually achieved adult dimensions.

The structure interposed between the less-developed face and the brain, the cranial base, is intermediate in its percent completion at birth (56 percent). Its rate of growth is more even than the calvarium when both the anterior (sella turcica–foramen caecum) and

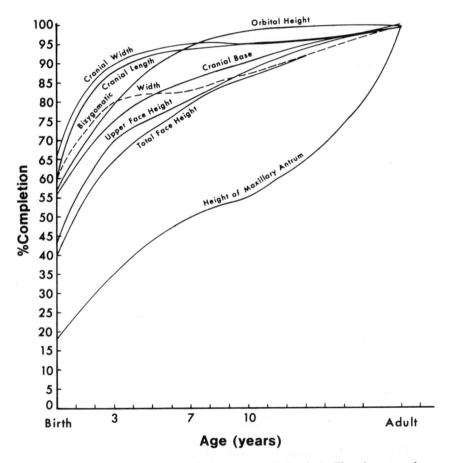

FIGURE 4.1 Cranial and facial measurements (males). The changing dimensions of the craniofacial complex. Data from J.H. Scott: The growth of the human face. Proc Roy Soc Med: 47:5, 1954.

the posterior (sella turcica–basion) segments are considered together, as in Fig. 4.1. Certain sutures of the cranial base undergo synostosis by the age seven or eight, a fact that has partly justified the use of the cranial base as reference plane in some cephalometric analyses.

The upper and the total face height are not half complete at birth (43 percent and 40 percent, respectively), which means that most striking and complex growth of the head is associated with the face. After an initial spurt during the first 3 years, the rate of increase of these dimensions slows but remains steady until the adult size is reached.

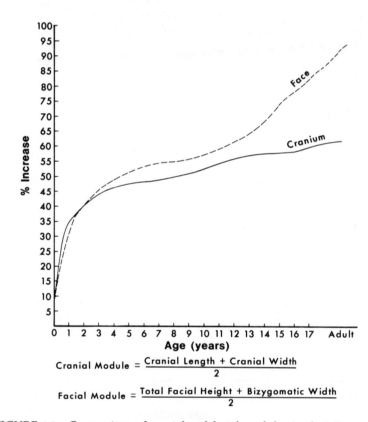

$$\text{Cranial Module} = \frac{\text{Cranial Length} + \text{Cranial Width}}{2}$$

$$\text{Facial Module} = \frac{\text{Total Facial Height} + \text{Bizygomatic Width}}{2}$$

FIGURE 4.2 Comparison of cranial and facial modules (males). Increase in cranial and facial modules during growth. Data from J.H. Scott: The growth of the human face. Proceedings of the Royal Society of Medicine 47:5, 1954.

Most of the increase in size of the maxillary antrum occurs postnatally (82 percent). The sinuses must wait for vertical maxillary and alveolar growth; the latter is dependent in turn upon the development and eruption of teeth.

Another useful depiction of the changing proportions of the head can be made by devising modules that average out two or more dimensions. For instance, a cranial module is created by adding the cranial length and cranial width and dividing by two. In Fig. 4.2, cranial and facial modules are plotted as percent increase over the birth size.

It can be seen that both areas expand slightly over 30 percent within the first year, after which the rate of growth slows; they continue to parallel each other for some time. The cranial module by adulthood has grown approximately 60 percent over its neonatal size. Near puberty, the facial module demonstrates a surge in growth, so that at maturity it has increased some 93 percent, almost doubling the dimensions recorded at birth.

ANATOMIC DIVISIONS

Introduction

The growth and development of the head is such an overwhelmingly complex subject when viewed as a whole, that we simply must disassemble it and work with the parts. Admittedly, this process will be quite arbitrary, and other viewers may see in the skull other division lines or natural separations or entities. Scott, for instance, visualizes nine craniofacial regions. However, most authorities on craniofacial growth would probably agree that the division of the head into three parts—neurocranium (the calvarium and cranial base), middle face, and lower face—heeds the dictates of anatomy and functions at least to the degree that any other arbitrary division would. Certainly, it allows careful scrutiny of important components of the skull without jeopardizing the integration of their growth processes. Lest the examination of the pieces of the puzzle prevent us from seeing the whole, Chapter 5 will be devoted to the woods and not the trees.

The Neurocranium

In man, the major portion of the skull is devoted to the housing of the brain. The calvarium and the cranial base together enclose, sup-

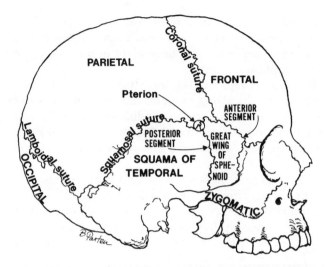

FIGURE 4.3 Lateral view of the skull emphasizing the complex suture system.

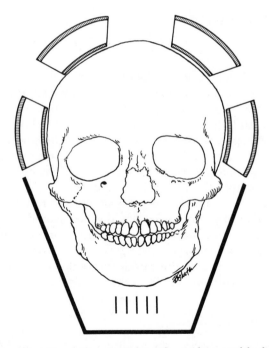

FIGURE 4.4 Growth of the cranial vault made possible by differential growth at the sutures. Adapted from M.L. Moss: The primacy of functional matrices in orofacial growth. Dent Practit 19:65, 1968.

port, and protect it, and for this reason, they will be discussed as a unit. Although the calvarium arises principally from membrane bone and the cranial base from the chondrocranium, they operate and grow in concert postnatally. The remainder of the skull—the middle face (maxilla, chiefly) and the lower face (mandible)—articulates with and is situated inferior to the cranium; from this position, the calvarium and cranial base set the pace for the other two regions and completely dominate the growth process. A reciprocal influence of the lower regions upon the cranium is virtually nonexistent.

The Calvarium

The calvarium consists of the following bones: an occipital, two parietals, the temporals, greater wings of the sphenoid, and a frontal. It has achieved approximately 63 percent of its growth at the time of birth and has completed 87 percent just 2 years later.

The cranial vault has two major suture systems that separate three cranial regions (Fig. 4.3). These are:

1. The coronal suture system, which above the pterion separates the frontal from the parietal bones and which below the pterion divides into an anterior part running between the frontal and great wing of the sphenoid and into a posterior segment running between the great wing of the sphenoid and the temporal. Pterion is the junction of the posterior end of the sphenoparietal suture and the squamosal suture.
2. The lambdoidal suture system, which passes between the occipital bone behind and the parietal and temporal bones anteriorly. These sutures create anterior, middle, and posterior cranial segments and converge at the spheno-occipital synchondrosis of the cranial base.

The calvarium and the cranial base grow in synchrony with the brain, and because this organ does not grow like a round balloon being filled, the covering bones are forced to exhibit a differential pattern of development. Although the capacity of the sutures to accommodate brain expansion is well illustrated in the simplified diagram of Fig. 4.4, the details of the complicated activity are not.

If we examine the brain diagramatically and rather simplistically, we find that it is a flexed structure composed of an inner zone (brain stem) and an outer zone (consisting chiefly of the cerebrum). The centripetally located brain tissue does not expand as much as

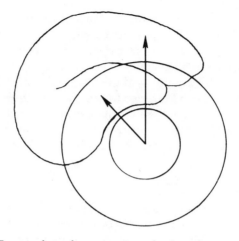

FIGURE 4.5 From a lateral perspective, the brain grows as if it were a
wedge from a circle. The portion closest to the center grows least,
whereas the farthest portion grows the most.

the outer zones, where extensive myelinization is occurring. One can
imagine a point positioned somewhere beneath the cranial base as
the center of two circles (Fig. 4.5), one congruent with the brain stem
(short arrow) and the other with the cerebrum (long arrow). The radius of the smaller circle will be found to increase at a lesser rate
than that of the larger circle. This is consistent with a pattern of
growth that maintains a flexure to the brain. Based on this fact, it
should become immediately apparent that sutures running from the
cranial base outward into the calvarium must demonstrate higher
growth rates peripherally.

Noting from Fig. 4.1 that cranial width and length are achieved
ahead of cranial base growth, we see that timing becomes a consider-
ation in the differential aspect of sutural development. If the cranial
base were more evenly paced with the calvarium, sutural growth at
area A would be more nearly comparable to that at area B (Fig. 4.6A).
Instead, calvarial expansion not only exceeds expansion of the cranial base, but it is manifested earlier. This additional factor helps to
force the tissues of the distal segments of the suture (D) to form at a
greater rate than the centrally located segments (C) (Fig. 4.6B). Thus,
while superimposed diagrams (Fig. 4.7) of the skulls of the same individual at different ages suggest that calvarial growth is a uniform
process, we must remember that it is the nonuniformity of sutural
apposition that makes such growth possible.

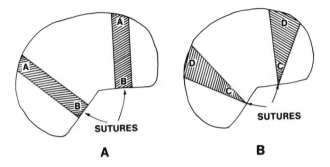

FIGURE 4.6 (A) The type of growth that would be exhibited in the cranial sutures of the brain stem and cerebrum expanded at the same rate and to the same extent. (B) Differential suture growth dictated by the growth pattern of the brain.

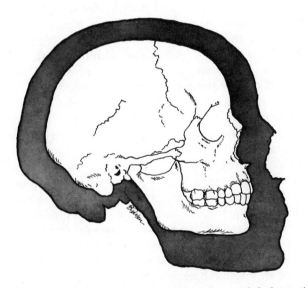

FIGURE 4.7 The apparent uniformity in cranial growth belying the differential growth that makes it possible.

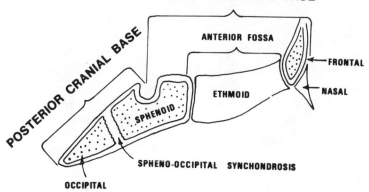

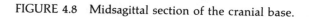

FIGURE 4.8   Midsagittal section of the cranial base.

The Cranial Base

DIMENSIONAL CHANGES WITH TIME

A midsagittal section of the skull reveals that complement of bones comprising the cranial base (Fig. 4.8). These bones and their lateral extensions house the anterior and middle brain segments; the important gland, the pituitary, is confined within the sella turcica of the sphenoid bone. The posterior cranial base runs from the posterior tip of the basioccipital (basion) to the sella turcica. The anterior cranial fossa encompasses the anterior section of the sphenoid and the ethmoid up to its junction with the frontal, a point accented by the foramen caecum. Normally, the clinical cephalometric anterior cranial base is measured from sella (S) to nasion (N).

Based on a study of dried skulls, Ford has suggested that the cranial base overall grows in an intermediate fashion between the growth of the cranium, which is dependent on the neural pattern of early and rapid growth, and the face, which conforms to a general skeletal pattern characterized by fairly even growth from birth until the pubertal spurt. He emphasized, however, that the individual parts of the cranial base manifest either a neural or general skeletal rate, but not an intermediate one. Between N and the anterior margin of the foramen magnum there is a region that exhibits a neural pattern, i.e., the one between the foramen caecum and S. The areas from N to foramen caecum and from S to basion (Ba) follow the general skeletal growth rate.

A clinical cephalometric study has confirmed that the anterior and posterior cranial bases do not exhibit similar growth patterns;

the anterior segment completes its growth several years before the posterior. Stamrud measured the anterior dimensions from many subjects of various ages and plotted them as a function of age (Fig. 4.9). What emerged from his study was a confirmation of previous assumptions—the anterior cranial base is a relatively slow-growing area that manifests an early completion. He determined that the sutural growth of the anterior cranial base (as represented by S–N less the thickness of the frontal bone) is finished around the eighth year. The distance S–N continues to increase for years after, but only as a result of apposition on the frontal bone.

As a result of this early completion, the anterior cranial base has been used wholly or in part as a reference in a number of cephalometric analyses. From a region of relative stability, the evaluation of other faster- and later-growing areas (the face) can be made. In contrast, the posterior cranial base is capable of linear expansion until adolescence, when the spheno-occipital synchondrosis ossifies.

RELATIONSHIP BETWEEN THE CRANIAL BASE AND CALVARIUM

The cranial base can increase in size anteroposteriorly after birth by growth at one synchondrosis and two sutures (Fig. 4.8). The synchondrosis lies between the occipital and the sphenoid; one suture lies between the ethmoid and the sphenoid, the other separates the ethmoid from the frontal. These bone growth mechanisms are not

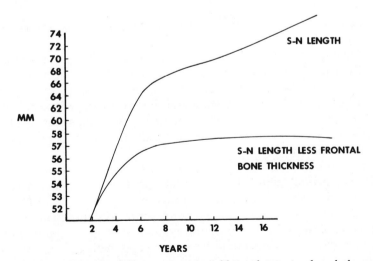

FIGURE 4.9   Growth of the anterior cranial base determined cephalometrically, with and without the frontal bone thickness. After L. Stramud: External and internal cranial base. Acta Odont Scand 17:239, 1959.

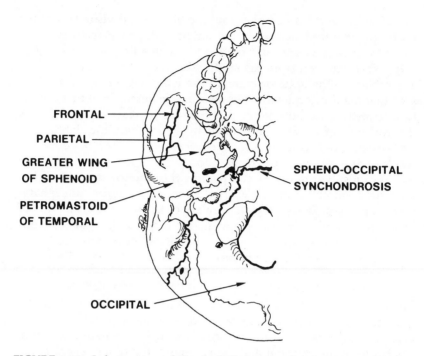

FRONTAL

PARIETAL

GREATER WING
OF SPHENOID

PETROMASTOID
OF TEMPORAL

SPHENO-OCCIPITAL
SYNCHONDROSIS

OCCIPITAL

FIGURE 4.10   Inferior view of the skull illustrating the convergence of cranial sutures on the spheno-occipital synchondrosis.

isolated anatomically; from two of them, sutures ramify across the
calvarium (Fig. 4.10).

Consider first the spheno-occipital synchondrosis. Spreading
laterally from it are two sutures, one sweeping back between the occipital and the petromastoid region of the temporal, the other
forward between the petromastoid and the great wing of the sphenoid (the latter part constitutes the posterior limb of the lower coronal
suture). The latter suture continues across the calvarium between the
frontal and parietals.

The sutures between the ethmoid and sphenoid travel out between the orbital plate of the frontal and the great wing of the
sphenoid to join the frontoparietal, or coronal, sutures (Fig. 4.11). In
addition, the anterior limb of the coronal suture system separates the
great wing of the sphenoid from the zygomatic bone within each orbital cavity, and at the cranial base, it merges with the suture between the small wing of the sphenoid and the frontal bone and
between the sphenoid and ethmoid bones. Thus, the anterior limb
separates the sphenoid bone and its various processes from the
frontal, ethmoid, zygomatic, and palatine bones.

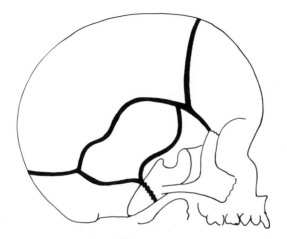

FIGURE 4.11 Lateral view of the skull depicting the convergence of the cranial sutures on the spheno-occipital synchondrosis and emphasizing the association between calvarial and cranial base growth.

Through these sutural links, the growth of the cranial base, the calvarium, and some of the face become closely interrelated. In particular, the spheno-occipital synchondrosis becomes a focus or pivot point of craniofacial growth. For this reason, it is difficult to imagine cranial base growth that would not have some impact on calvarial or facial sutures.

REMODELING OF THE CALVARIUM

The apposition–resorption pattern of the cranium might surprise the novice craniofacial biologist. Not unexpectedly, the entire outer surface of the calvarium is depository; apparently the bone drift outward in contributing to cranial expansion. Quite unexpectedly, however, most of the calvarium is also depository on the inner surface (Fig. 4.12). At times, resorption is observed on the inner surface, because during remodeling the concavity of the cranial bones must be reduced. Nevertheless, the net pattern is depository, and as a result, the bones increase in thickness during growth.

The demarcation between apposition and resorption is quite abrupt. This phenomenon is easily understood when the anatomy of the cranial floor is examined. Because the lobes of the brain expand within bony fossae that are limited in their ability to respond by sutural adaptation, the only recourse is bone removal (Fig. 4.13). On the other hand the vault of the cranium can be appositional on both sides of the bones, because sutural response alone is capable of accommodating brain expansion.

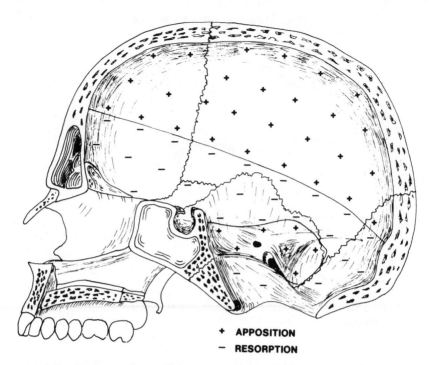

+ **APPOSITION**

− **RESORPTION**

FIGURE 4.12   Interior view of the cranial vault illustrating the demarcation between apposition and resorption. Adapted from R.E. Moyers: Handbook of Orthodontics, Chicago, Year Book Medical Publishers, Copyright © 1972.

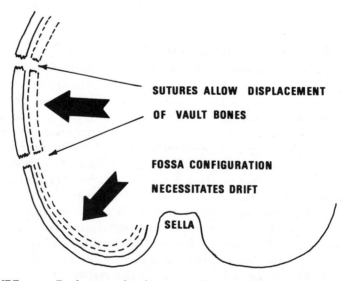

SUTURES ALLOW DISPLACEMENT OF VAULT BONES

FOSSA CONFIGURATION NECESSITATES DRIFT

SELLA

FIGURE 4.13   Explanation for the reversal between apposition and resorption in the cranial vault. After D.H. Enlow: Handbook of Facial Growth. Philadelphia, Saunders, 1975.

Thus, at first glance the growth of the cranium might seem a straightforward process, but on close inspection, the adaptive sutural growth closely attuned to cranial base growth and the unexpected apposition and resorption pattern quickly dispel this notion. We can soon appreciate that the potential for differential growth of the whole system is a stunning example of an evolutionary accommodation.

The Middle Face

The middle face consists of the orbits and their contents, the nasal cavity, the maxillary sinuses, and the upper alveolar process and teeth. The primary bone is the maxilla, but contributions are made by the nasal, frontal, ethmoid, vomer, lacrimals, conchae, palatine, and zygomatics.

On superficial inspection, all the sutures that attach the middle complex of bones to the cranium, i.e., frontomaxillary, zygomatico-maxillary, zygomaticotemporal, and pterygopalatine, are aligned in such a way that any growth there would project the middle face downward (Fig. 4.14). In this manner, the middle face could maintain

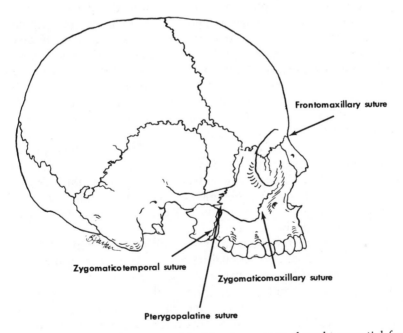

FIGURE 4.14 The alignment of facial sutures once thought essential for the downward and forward growth of the maxilla.

its anteroposterior relationship to a growing anterior cranial base and at the same time increase its vertical dimension.

For some number of years, middle face growth was explained by this sutural theory, but the emergence of new concepts in recent years has virtually discounted it. Two lines of study made the suture theory untenable. First, no one could demonstrate autonomous expansive growth of a suture; that is, sutures were found to fill in only as other forces separated bones. Second, the actual alignments of the various sutures are not uniformly downward and forward, and only by proposing unusual attributes of cell activity could any unidirectional growth be realized. Coben views these sutures as "growth adjusters" and not as "growth initiators." Although a controversy still rages over what does provide impetus for midfacial growth, most authorities would assign only a secondary function to sutures.

Dimensional Changes

The midface of the growing child undergoes a dramatic increase in absolute terms, but particularly in relation to a cranium that is more nearly complete at birth. The middle face expands in width, depth, and height, but the last dimension is most remarkable.

Scott views the sphenoethmoidal suture, the palatomaxillary suture, and the pterygopalatine suture as part of one integral circummaxillary suture system that permits the nasal cartilage to drive the maxillary complex away from the sphenoid bone, opening space for the erupting posterior teeth. He considers that this entire system ceases to grow after the closure of the sphenoethmoidal suture at the age of seven. According to Scott, the growth of the middle face falls into two distinct phases: (1) from birth to about the seventh year of age and (2) after the seventh year.

During the first seven years, brain expansion lengthens the anterior cranial base, growth of the eye expands the orbital cavity, and the nasal cartilage thrusts the maxilla downward. The other cartilage replacement mechanisms, the spheno-occipital synchondrosis, and condylar cartilage are concurrently active; surface apposition makes only a minor contribution (Fig. 4.15).

After the seventh year, the brain and éyes are virtually complete, so that the anterior cranial base, as measured from the pituitary fossa to the foramen caecum, ceases growth and the orbital cavities stabilize. Simultaneously, the nasal cartilage ceases to grow. The slowdown in all these growth mechanisms effects a cessation of sutural activity. In this phase of growth, the increase in size of the middle face results mainly from surface deposition and internal reconstruction (Fig. 4.16).

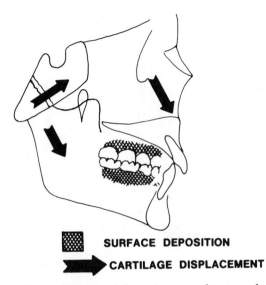

SURFACE DEPOSITION

CARTILAGE DISPLACEMENT

FIGURE 4.15 Craniofacial growth systems to the age of seven, surface deposition contributing only in a minor way. Arrows depict cartilage growth. Adapted from S.E. Coben: Growth and class II treatment. Am J Orthod 52:5, 1966.

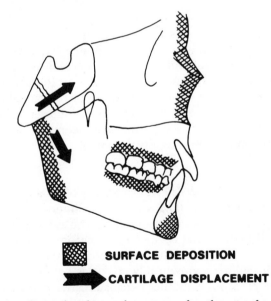

SURFACE DEPOSITION

CARTILAGE DISPLACEMENT

FIGURE 4.16 Craniofacial growth systems after the age of seven, illustrating the significant role played by apposition. Arrows represent cartilage growth. Adapted from S. E. Coben: Growth and class II treatment. Am J Orthod 52:5, 1966.

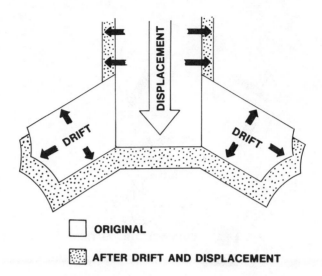

ORIGINAL

AFTER DRIFT AND DISPLACEMENT

FIGURE 4.17   Frontal diagram of growth of the nasal cavity and sinuses illustrating the roles played by drift and displacement. Adapted from
R.E. Moyers: Handbook of Orthodontics. Chicago, Year Book Medical Publishers, Copyright © 1972.

Figure 4.17 depicts the overall changes in the maxilla from a
frontal perspective. The nasal cavity enlarges laterally through the
process of drift, while the large vertical change is wrought by a combination of displacement and drift. The process of resorption on the
floor of the nose in concert with apposition on the palate abets the
displacement activity of the nasal cartilage or functional matrix (see
Chapter 6). Drift increases the volume of the sinuses superiorly, inferiorly and laterally, but a superior encroachment into the orbital
cavity is nullified by opposing displacement.

GROWTH IN WIDTH

Comparing the skull of the newborn to that of an adult, an obvious
lateral expansion is seen (Fig. 4.18). The dimension A, or cranial
width (refer to Fig. 4.1), increases rapidly. The eyes, also early to develop, expand their orbital cavities in synchrony with the widening
cranial floor. Interestingly, the interorbital distance (B) increases
very little. Bizygomatic growth (Fig. 4.1) lags behind cranial growth
in the earlier years but continues expansion after cranial activity has
leveled off; as a result, the cheek bones tend to become relatively
more prominent with age.

The principles of bone growth responsible for the lateral growth
of the face involve displacement and drift, the former dominant

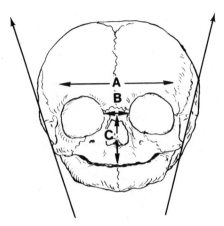

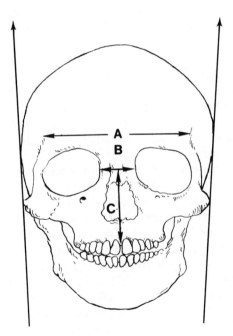

FIGURE 4.18　Frontal view of the skull of the newborn compared to that of an adult, showing important dimensional changes. After D.H. Enlow: *The Human Face*. New York, Harper and Row, 1968.

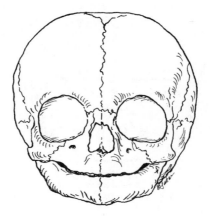

FIGURE 4.19   Frontal view of a fetal skull illustrating the anterior suture system, which allows lateral displacement growth. Adapted from J.H. Scott: The growth in width of the facial skeleton. Am J Orthod 43:366, 1957.

during the early stages of growth and the latter during the final stages. The changeover from displacement to drift reflects the maturation of the embryologic developmental process.

At birth, the skull shows the presence of a complete sagittal suture dividing the cranium into two halves. The external suture system is composed of (1) the metopic suture separating the frontal bone; (2) the internasal suture between the nasal bones; (3) the intermaxillary suture, which extends back along the middle of the hard palate as the midpalatal suture; and (4) the mandibular symphysis (Fig. 4.19). The interior of the skull is not bisected cleanly but is complicated by a midline cranial base (Fig. 4.20).

From the foramen magnum to the foramen caecum (anterior end of the mesethmoid), the midline cranial base is preformed in cartilage and later develops the following ossification centers: basioccipital, postsphenoid, presphenoid, and mesethmoid. At birth, the post- and presphenoid centers have united to form the body of the sphenoid, which still remains separated from the basioccipital by the spheno-occipital synchondrosis. The mesethmoid commences to ossify shortly after birth.

The interior suture system divides to run on either side of the middle cranial base and is made up of the following parts, front to back:

1. the metopic frontal suture to the foramen caecum;
2. the cribriform plate;

3. the body of the sphenoid, which forms the medial walls of each orbital cavity, situated beneath the laterally spreading lesser wings;
4. the greater wings of the sphenoid (membrane bone) separated from the body of the sphenoid by an area of cartilage; and
5. the petrous parts of the temporal bones separated from the side of the body of the sphenoid and from the occipital bone by connective tissue and the jugular foramen.

Thus at the time of birth, the sagittal suture system is capable of responding to (1) brain growth and (2) growth of cartilage. However, during the first year, important changes occur that quickly diminish the role of this sutural system in lateral growth. The frontal bone and the mandible become continuous bones, as the metopic suture of the former and the symphysis of the latter unite. Concurrently, the greater wings become united to the body of the sphenoid, with

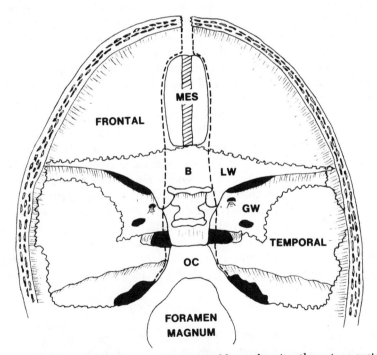

FIGURE 4.20 Neonatal state of the cranial base showing the suture system which can accommodate lateral brain expansion; Mes, mesethmoid; B, body of sphenoid; Oc, occipital; LW, lesser wing of sphenoid; GW, greater wing of sphenoid. Adapted from J.H. Scott: The growth in width of the facial skeleton. Am J Orthod 43:366, 1957.

the result that a single bone extends from one temporal fossa to the other. By the third year, the cribriform plate ossifies, uniting the ethmoid to the mesethmoid (perpendicular plate), so that a single bony unit extends from one orbital cavity to the other.

As a consequence of this sagittal system, early lateral brain expansion can be accommodated by separation of the parts of the cranial floor. Once the sphenoid and its wings are joined, drift must be held responsible for any subsequent increase in cranial width.

By about 3 years of age, the distance between the eyeballs has achieved adult dimensions, a process coincident with the completion of the ethmoid complex. From this time on, increase in the lateral orbital dimensions results from growth of the eyeballs. Part of the space for the eyes is probably achieved by outward growth of the zygomatic bones.

As the interocular distance is increasing during the early years, the maxilla is able to follow by its expansion capabilities at the midpalatal suture. Unlike the bones above, which are more closely related to cranial base, the maxilla maintains some capacity to expand beyond the time of fixation of the sagittal sutural system. With the completion of orbital growth around the seventh year, the middle face becomes stabilized, and nasal cavity expansion by displacement ceases. Further enlargement involves a surface deposition and internal resorption.

The distance between the facial buttresses continues until adulthood and can be ascribed to lateral displacement and drift of the cheek bones. The displacement of the zygomatic process laterally increases the space for the temporalis muscle and maintains a horizontal relationship with an ever widening mandibular ramus. Drift accommodates the volumetric expansion of the maxillary antra.

Because the medial interorbital distance changes little, the lateral interorbital increase reflects essentially the enlargement of the eyes, a process complete early in life. As a result, the later-expanding nasal cavity, sinuses, and zygomatic processes surpass the orbital area and cranium in growth, and the facial proportions are altered (Fig. 4.18).

GROWTH IN HEIGHT

Compared to dimension B of Fig. 4.18, the vertical increase (C) is striking, doubling during growth. A part of this gain in facial height can be ascribed to the alveolar process, but most is due to an enlargement of the nasal cavity. At birth the floor of this cavity is at the level of the inferior margins of the orbits, but there is a substantial separation of these landmarks with time. Evidently, the assurance of an adequate nasal capacity is a fundamental drive in facial growth; as the rest of the body grows and increases its respiratory demands, the

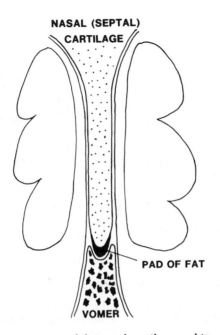

NASAL (SEPTAL)
CARTILAGE

PAD OF FAT

VOMER

FIGURE 4.21   Cross section of the nasal cartilage and its relationship to the vomer. Adapted from J.H. Scott: Dento-facial Development and Growth. Oxford, England, Pergamon Press, 1967.

face must be able to respond in kind. If you do not consider this an evolutionary prerequisite, remember how miserable life became during your last episode of nasal congestion.

Roughly half of the vertical growth results from displacement, the other half from drift. Remember, however, that during the early stages, the two activities are occurring simultaneously. The tissue(s) that displaces the maxilla downward has not yet been definitively affixed to everyone's satisfaction. Two conflicting proposals to explain this process are the nasal cartilage theory and the functional matrix theory.

The potential for the cartilage of the nasal cavity to act as a growth center has been previously described. According to the nasal cartilage theory, the cells of the cartilage are genetically programmed to divide and synthesize in a manner that accommodates the needed capacity for air exchange. The nasal cartilage thrusts the maxilla downward, separating those "aligned" sutures, which then respond by deposition.

At birth, the nasal septum, consisting of cartilage, is continuous with the cartilage of the cranial base. Inferiorly, most of the septum sits in the vomerine groove, which by virtue of a mass of loose fatty tissue, prevents fusion between the septum and vomer (Fig. 4.21). In

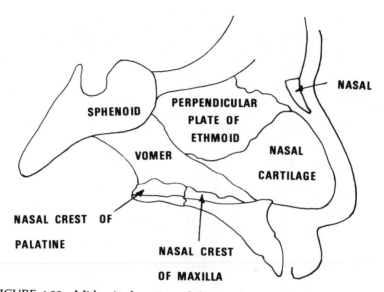

FIGURE 4.22 Midsagittal section of the nasal cavity illustrating the relationship between the nasal septum and the maxilla.

front the septal cartilage is firmly united by fibrous tissue to the premaxilla in the region above and behind the nasal spine of the nasal aperture. Nature has thus provided an arrangement that best utilizes the separating force of the septum while allowing for the maximum adjustments along the groove of the vomer (Fig. 4.22).

Toward the seventh year, the perpendicular plate of the ethmoid (the ossified nasal cartilage) unites with the vomer behind the septal cartilage. Presumably, this marks the end of the displacement period of the maxilla, although growth of the septal cartilage could theoretically separate the maxillary and palatine bones from the ethmoid and vomer beyond this time.

The alternative theory, the functional matrix theory, about which more will be said in a later section, suggests that the nasal cartilage plays no greater role than the sutures. What really displaces the maxilla is the sum total of the muscles, tissues, nerves, and function surrounding the bone and cartilage; no single entity, such as nasal cartilage, has the capacity to move a complex structure like the middle face.

Regardless how this "why" controversy is resolved in ensuing years, continue to bear in mind the reality of displacement in this region and in others as well.

Drift accounts for about half of the vertical growth in the midface. The volume of the nasal cavity is increased laterally and inferiorly by resorption of the internal faces of the surrounding bones.

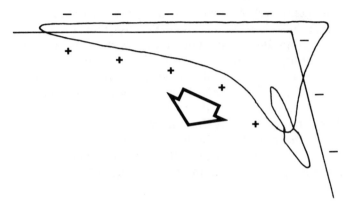

FIGURE 4.23   Midsagittal section of the maxilla. Growth visualized as following the V principle. After D.H. Enlow and S. Bang: Growth and remodeling of the human maxilla. Am J Orthod 51:446, 1965.

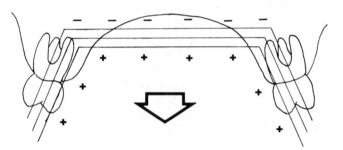

FIGURE 4.24   Cross section of the maxilla. Growth visualized as following a V, albeit somewhat modified. After D.H. Enlow and S. Bang: Growth and remodeling of the human maxilla. Am J Orthod 51:446, 1965.

To prevent perforations, it is obvious that the opposite surfaces of these bones must be undergoing deposition.

The activity involved in palatal drift is a classic example of drift as a growth mechanism. The nasal floor is resorptive in nature, while the opposite surface, the roof of the mouth, displays bone formation. Some of the patterns of deposition–resorption in the maxilla can be illuminated by judicious use of the V principle. If the midsagittal section of the maxilla is displayed as a shallow V (Fig. 4.23), the inside or palatal side is found to be depository, while the floor of the nasal cavity and the surface just above the maxillary incisors (both of which are on the side opposite to the direction of growth) are resorptive. A cross section of the palate discloses another V and quickly identifies the areas of apposition and resorption (Fig. 4.24). The net result of this drift is a movement of the palate away from the cranial base and an enlargement of the nasal cavity.

The resorptive phenomenon over the maxillary incisors required that they change their angulation lest their roots become exposed. Appropriately, the roots move posteriorly and the teeth become steadily less upright. This teeth movement blends harmoniously with the growth of the lower face and teeth as the profile changes from a retrogothic to an orthognathic configuration. (If the process is excessive, prognathism results.)

Do not be confused by the resorptive nature of the anterior surface of the maxilla—the middle face is not becoming diminutive. Displacement is actually carrying the maxilla forward; drift is simply not contributing and is, in fact, marginally combatting the movement.

GROWTH IN DEPTH

We have just alluded to the anterior displacement of the maxilla as it maintains a relationship with the forward-growing anterior cranial base. The maxilla does not grow forward or make room for erupting teeth by depositing new bone on the anterior surface. Instead, new growth occurs posteriorly in the maxillary tuberosity area. As the first permanent molars, then the seconds, and finally the thirds are preparing to move toward occlusion, room must be made. New space is developed by displacement of the whole maxilla forward followed by growth from behind. In effect, the maxilla is growing posteriorly but is being displaced anteriorly. In this way, some minor resorption can occur on the anterior surface of bones despite their net forward movement.

The maxilla does not actually grow against the pterygoid plates in such a fashion that it thrusts itself forward; bone is much too pliable in that sort of environment to withstand the pressures. Instead, the nasal cartilage, or some other force, is creating a space into which bone and erupting teeth can develop. If there is enough anterior displacement, then all the posterior teeth, including the wisdom teeth, can erupt satisfactorily. Apparently this occurs only in a segment of our population, judging from the number of surgical interventions necessary. The corner buttresses of the face, the zygomatic processes of the maxilla, are naturally carried forward with the rest of the complex, adjustment made possible by the zygomaticomaxillary sutures. However, one can observe in the adult that these processes are not situated as far anteriorly with respect to the nasal area as they are in the child. Obviously, during the displacement process, the zygomatic areas are lagging behind, a process attributable only to a reverse drift. In Fig. 4.25 it can be visualized that the zygomatic process remains relatively fixed in space, while the remainder of the maxilla is thrust forward; this can only be achieved by resorption anteriorly and deposition posteriorly.

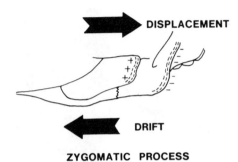

FIGURE 4.25   The process of anterior resorption and posterior deposition that prevents the zygomatic process from being displaced forward to the same extent as the rest of the face. After D.H. Enlow: The Human Face. New York, Harper and Row, 1968.

The resorption in this anterior region, combined with the aforementioned resorption over the central incisors, means that the increase in depth of the middle face is accomplished by posterior growth–anterior displacement, with drift contributing minimally.

In a nutshell, the middle face can be described as growing downward and forward, the former direction most prominent, the latter just sufficient to maintain its position under the cranial base. This growth should be contrasted with that of the lower face, which not only grows vertically in dramatic proportions, but swings forward from a retracted position into a vertically balanced alignment with the cranial base.

The Lower Face—The Mandible

The lower face consists of the soft tissues attached to one complex bone, the mandible. Some years ago, a standard textbook described the mandible as a long bone bent in the shape of a U with cartilage growth plates at each end that caused growth and offered articulation. The endochondral growth mechanism was stressed as the major contributor to mandibular formation. In recent years, the role of the condylar cartilage in overall mandibular development has been minimized, its function restricted to condylar development and articulation. Animal research and human clinical studies have determined that the mandible attains an almost normal size and position in space exclusive of condyles. When this occurs, however, there is an alteration in mandibular shape, because extraordinary muscular forces must be generated to compensate for the lack of fulcrum points.

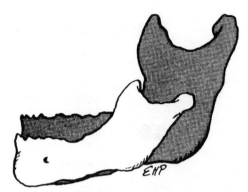

FIGURE 4.26 Serial pictures of a mandible superimposed on the symphysis, illustrating that growth occurred posteriorly.

Posterior Growth–Anterior Displacement

The mandible is an excellent example of posterior growth–anterior displacement. After the arch is formed in utero, the only significant contributions to overall size by anterior apposition occur within the first few months of life. After this time, outside of symphyseal growth and minor deposition on the alveolar process during tooth eruption, all of the forward movement of the mandible is accounted for by growth from behind.

When likenesses of the mandible of one individual at two different ages are superimposed on reference points (these can be embedded metallic implants or certain anatomic landmarks), the extent of the posterior growth becomes evident (Fig. 4.26). Room for the three permanent molars must be obtained, not by forming new bone as in the maxilla, but by resorbing the leading edge of the ramus. Obviously if this process were not balanced by deposition on the posterior surface, the ramus would narrow. Actually, there is substantial apposition, which is synchronized with cartilage replacement activity in the condyle. Since the condyle is buttressed against the glenoid fossa, its posterior growth is translated into anterior displacement.

Vertical Growth

Interocclusal vertical dimension has been shown to be one of nature's constants; that is, each individual maintains virtually the same distance between his teeth (lower jaw in rest position) throughout his life.

Not only must the mandible increase its vertical dimension substantially to insure that the erupting teeth never encroach upon this zone, but it must also compensate for the posteruption vertical movement of teeth, the increase in height of the alveolar processes, and for the entire descent of the middle face. This means that the mandible must grow substantially just to stay even.

In addition to this burden, the profile of a growing child characteristically becomes less retruded. This implies an anterior component to mandibular growth that must not only be sufficient to keep up with maxillary growth but must also surpass it.

Because of the intimate relationship between the middle face and the cranial base and the more remote relationship between the mandible and cranial base, the horizontal position of the maxilla is seldom contributory to faulty anteroposterior relationships between the jaws. The substantial growth required of the mandible combined with its minimal articulation with other bones makes it more vulnerable to aberrant conditions.

Patterns of Apposition and Resorption

Enlow has pointed out that the sites of bone apposition and resorption in the mandible have little to do with the insertion of muscles. Figure 4.27 compares the areas of bone formation with the attachment of muscles. Evidently, the contraction of muscles does not have an exclusively localized influence on their periosteal insertions. As

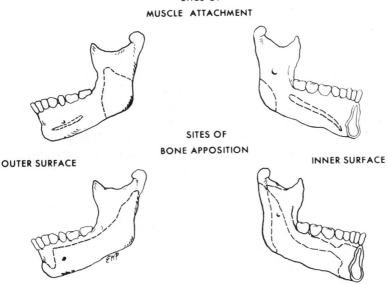

FIGURE 4.27 A diagram emphasizing the complexity of the relationship between muscle forces and bone response; sites of muscle attachment are not necessarily correlated with bone apposition.

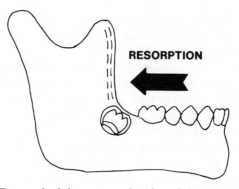

FIGURE 4.28   Removal of the anterior border of the ramus to make room
    for erupting molars.

discussed in Chapter 3, the growth in some areas of the mandible
can be explained by the law of electrogenesis, which describes the
fact that a load tends to make a surface more concave or more convex, leading to apposition or resorption, respectively. Whether its
concept is valid, or whether the pattern is a product of genetic determination, a manifestation of the functional matrix (see Chapter 6), or
a phenomenon of something yet untheorized remains to be
answered.

THE RAMUS AND CORONOID PROCESS

Speculation aside, even a superficial examination of the apposition–resorption patterns tells us that a more complicated scheme is
involved than the simple posterior addition of new bone on the rami.
It also becomes abundantly clear that a simplistic drawing of ramal
resorption (Fig. 4.28) where the anterior border is removed to provide room for erupting molars conveys what is happening only in
very general terms. While it is true that the leading edge of the
ramus drifts posteriorly, the position of the coronoid process (the
inner surface faces medially, superiorly, and anteriorly) dictates that
surface rather than an edge is involved.

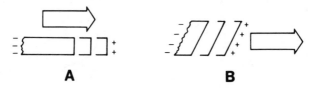

A   **B**

FIGURE 4.29   (A) Drift of bone by edge apposition and resorption. (B)
    Drift of bone by surface apposition and resorption. After D.H.
    Enlow: The Human Face. New York, Harper and Row, 1968.

Figure 4.29 illustrates the drift of a flat bone by two processes, (1) edge apposition and resorption and (2) whole surface involvement. The drift of the coronoid process is an example of the latter type.

To understand how this can be accomplished let us return to the V principle. Taking note of the cross section through the coronoid processes in Fig. 4.30, we see a configuration that is very suggestive of a long bone previously discussed. The coronoid processes are an example of an ever-expanding V with the inside surface undergoing apposition and the outer surface resorption. Enlow describes these processes as having propellerlike cant, which precludes the straightforward resorption of the anterior border.

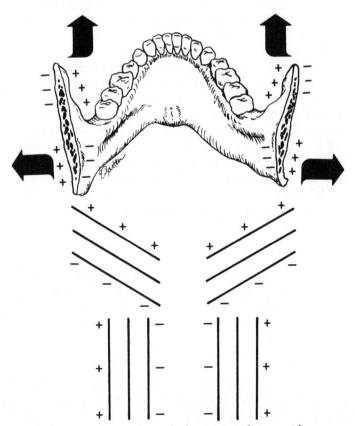

FIGURE 4.30   Cross section through the rami and coronoid processes, illustrating how these areas of the mandible can be visualized as following the V principle. Adapted from D.H. Enlow and D.B. Harris: A study of the postnatal growth of the human mandible. Am J Orthod 50:25, 1964.

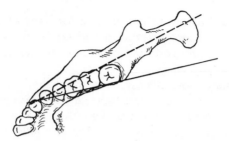

FIGURE 4.31 Superior view of the mandible. The teeth do not diverge with the body of the mandible, but remain closer to the midline.

Further applications of the V explain the apposition on the buccal surface of the ramus and the resorption of the lingual surface beneath the mylohyoid ridge. These combined activities profoundly influence the shape of the mandible in two ways. First, the lateral ramus deposition is part of the drift that leads to the squarer jaw of the adult. Second, when one looks at the mandible from above (Fig. 4.31), it is apparent that the process is designed to keep the teeth closer to the midline than are either the rami or the coronoid processes. In fact, the position of these lower teeth is virtually inviolate—orthodontists have met with little success in attempting to expand the lower arch. Eons ago it was decided, for some reason, that the lower teeth would not follow the basal portion of the mandible as in other mammals, but would remain in alveolar bone nearer the midline. The resulting anatomic configuration has consistently plagued the dentist involved in the construction of artificial dentures.

THE CONDYLE

Each individual condyle is also remodeled according to the V principle. In Fig. 4.32, the condylar cartilage can be visualized as being situated on the inside of an ever-expanding V with its neck lengthened by reduction of the bone situated on the side of the V away from the direction of growth.

THE CORPUS

While not as active as the coronoid and condylar processes, the basal portion of the mandible undergoes an interesting reversal in its apposition–resorption pattern from the posterior aspect toward the anterior (Fig. 4.33). At the ramus, the buccal surface is depository and the lingual surface is resorptive, while anteriorly, these fields have tended to reverse themselves, with the result that the symphysis is completely appositional on the lingual but only partially so at the

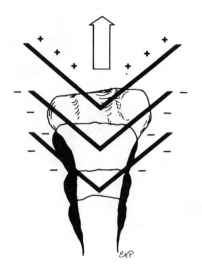

FIGURE 4.32   Growth at the condyles following the V principle.

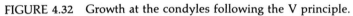

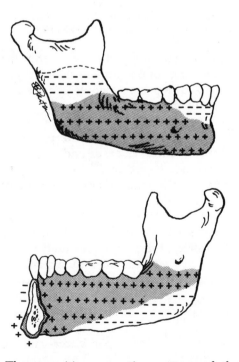

FIGURE 4.33   The apposition–resorption pattern of the body of the mandible.

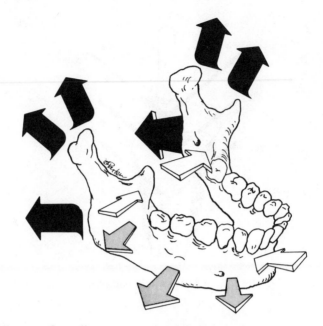

FIGURE 4.34   Overall summary of the growth of the mandible. Adapted
from D.H. Enlow and D.B. Harris: A study of the postnatal growth of
the human mandible. Am J Orthod 50:25, 1964.

chin point. The alveolar bone immediately superior to this chin point
is actually resorptive. The combination of these activities at the front
of the mandible means that most people acquire a prominence to
their chins during growth.

Resorption of the anterior plate of alveolar bone necessitates a
continual uprighting of the lower incisors to prevent root exposure.
Fortunately, this movement does occur and in such a way that it is
coordinated with the downward and forward displacement of the
mandible and the change in long axis of the maxillary incisors.

OVERALL VIEW

Figure 4.34 summarizes the overall growth of the mandible. The
ramus grows posteriorly, at the same time increasing the interramal
dimension in synchrony with the laterally expanding cranial floor.
By virtue of V principle growth, the coronoid processes grow superiorly and buccally, the anterior edges of the rami being continually resorbed in the process. The condyles are growing posteriorly,
superiorly, and laterally.

Apposition is characteristic of the buccal surface of the lower
part of the ramus, the inner surface of the coronoid and condylar

processes, and most of the buccal surface of the basal portion. Resorption is prominent under the mylohyoid ridge, laterally on the superior aspect of the ramus and its processes, and inferior to the anterior teeth.

Further discussion of mandibular growth will be postponed until Chapter 7, at which point the background material presented will facilitate completion of the subject.

The Frontal Face

From a frontal view, the contrasting of apposition and resorption of the entire face is very illuminating. The majority of the external surfaces of the bones in the head are depository; those areas associated with the anterior dentition and anterior zygomatic processes are resorptive (Fig. 4.35).

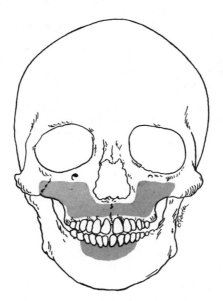

FIGURE 4.35   Frontal view of the human skull delineating the areas of apposition (nonstippled) and resorption (stippled). Adapted from D.H. Enlow: The Human Face. New York, Harper and Row, 1968.

BIBLIOGRAPHY

Bjork A: Sutural growth of the upper face studied by the metallic implant method. Acta Odontol Scand 24:109, 1966
Brodie AG: On the growth pattern of the human head. Am J Anat 68:209, 1941

Brodie AG: The behavior of the cranial base and its components as revealed by serial cephalometric roentgenograms. Angle Orthod 25:148, 1955

Coben SE: Growth and class II treatment. Am J Orthod 52:5, 1966

Diamond M: Posterior growth of maxilla. Am J Orthod 32:359, 1946

Enlow DH: The Human Face: An Account of the Postnatal Growth and Development of the Craniofacial Skeleton. New York, Harper and Row, 1968

Enlow DH: Wolff's law and the factor of architectonic circumstance. Am J Orthod 54:803, 1968

Enlow DH: Facial Growth and Development. In Moyers RE (ed): Handbook of Orthodontics, 3rd ed. Chicago, Year Book Medical, 1972

Enlow DH, Bang S: Growth and remodeling of the human maxilla. Am J Orthod 51:446, 1965

Enlow DH, Harris DB: A study of the postnatal growth of the human mandible. Am J Orthod 50:25, 1964

Enlow DH, Hunter WS: A differential analysis of sutural and remodeling growth in the human face. Am J Orthod 52:823, 1966

Ford EHR: Growth of the human cranial base. Am J Orthod 44:498, 1958

Goldstein MS: Changes in dimension and form of the face and head with age. Am J Phys Anthropol 22:37, 1936

Greenberg A: Life cycle of the human mandible. NY Dent J 31:98, 1965

Latham RA, Burston WR: The postnatal pattern of growth at the sutures of the human skull. Dent Pract 17:61, 1966

Powell TW, Brodie AG: Laminagraphic study of the spheno-occipital synchondrosis. Anat Rec 147:15, 1963

Roche AF: Increase in cranial thickness during growth. Hum Biol 25:81, 1953

Scott JH: The cartilage of the nasal septum. Br Dent J 95:37, 1953

Scott JH: The growth of the human face. Proc R Soc Med 47:91, 1954

Scott JH: Craniofacial regions: Contributions to the study of facial grwoth. Dent Pract 5:208, 1955

Scott JH: The growth in width of the facial skeleton. Am J Orthod 43:366, 1957

Sicher H: The growth of the mandible. Am J Orthod 33:30, 1947

Stramud L: External and internal cranial base: A cross-sectional study of growth and association in form. Acta Odont Scand 17:239, 1959

· 5 ·

Integration of
Craniofacial Growth

The magic of a face.
Thomas Carew

SUPERIMPOSITIONING

The Problems

The last chapter examined the components of facial growth individually; this chapter will attempt to integrate these divisions and explore their balance and harmony.

Traditionally, the entire facial complex of an individual is analyzed for longitudinal growth by superimposition of lateral head film tracings. In this technique, serial cephalograms of one individual are taken and traced on transparent acetate sheets. Then the tracings are placed one over the other, superimposing certain landmarks or planes drawn between landmarks, in order to obtain an overall impression of growth. Unfortunately, there are a thousand landmarks and planes from which to choose, and the selection of those most suitable remains a nagging problem to researchers and clinicians.

To better understand the dilemma facing the investigator of longitudinal facial growth, consider the use of standardized serial photographs for evaluating the changes in body height, proportions, and posture of children. In this procedure, the subject stands on a designated spot, usually close to a wall, with a camera positioned a measured distance away; lighting, lens settings, and background are fixed. Photographs taken in such a setup, with all the parameters held constant year after year, present to the student of development comparably scaled visual images of growing children. These pictures can be aligned side by side, an arrangement that locates the feet of the subject on the same horizontal plane. Differences in height between certain anatomic landmarks and the floor can now be easily seen and documented.

In this kind of longitudinal somatic growth study, the floor is the reference plane; it is an absolute. Unfortunately for craniofacial biologists, there is no such stable reference area in the head. If the floor were uneven or the camera position constantly changing, then meaningful comparisons of body growth would be difficult. Because of regions of the head are changing, there is virtually no way to be absolutely certain of any landmarks or judgements made about growth based on these landmarks.

To circumvent this problem, acceptable compromises have been made. Areas of relative stability are used as points of reference to assay growth in other faster and/or longer-growing areas. For example, the cranial base, particularly the anterior segment, is often used to analyze the positions of the maxilla and mandible.

Because the cranial base is a natural line of demarcation between the cranium and the face and because some of its components cease growth early relative to the rest of the head, it has become a reference structure for static and dynamic cephalometric analyses.

In the development of the cranial base as a registration point for superimposition in longitudinal studies, it soon became apparent that the structure is not so limited that a number of positions cannot be visualized. Depending upon the interest and awareness of the observer, any number of orientations can be used that will yield invaluable information.

It must be pointed out, however, that the way one superimposes serial tracings on the cranial base influences the position of the face in space and may add to or subtract from real growth.

S-N Plane

Consider one of the easiest and most popular techniques for superimposition, the S-N plane. In Fig. 5.1, the tracings are positioned

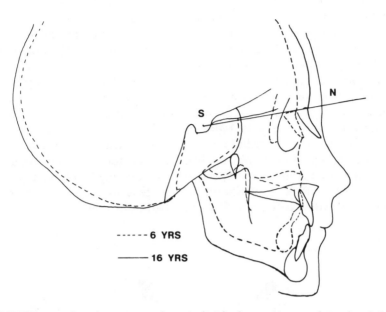

FIGURE 5.1 Serial tracings of an individual superimposed in the S–N plane with sella as the reference point.

on the center of the sella turcica (S) and aligned on the line drawn between sella and nasion (N), the junction between the frontal and nasal bones. From this advantage point, a definite feel for the flow of growth is tendered, but unfortunately, some distortion has been introduced. Let us see why.

If the growth of the head is analyzed from its position on the cervical vertebrae (Fig. 5.2), it should be immediately obvious that there are not only many examples of posterior growth–anterior displacement but also of inferior growth–superior displacement. For example, the bone forming activity of the spheno-occipital synchondrosis is directed inferiorly and posteriorly against the atlas, the first cervical vertebra. Since this vertebra is not resorbed during the process, the only result can be displacement of the whole head upward. Thus, in space, part of the cranial base changes position just as other areas of the head do. Additionally, the sphenoid bone can undergo superior bone apposition.

If what you want to do is determine the extent of mandibular growth relative to the cranial base, then a superimposition on the cranial base is legitimate. If on the other hand, you wish to describe mandibular growth as it is really occurring in space, then you must take into account that the cranial base is also involved in displacements and apposition. Probably, the orientation of longitudinal cephalometric tracings that most reflects the true growth of the head in space has been suggested by Coben. He superimposed on the basion and paralleled the anterior cranial bases. By this technique, the extent of craniofacial growth away from the cervical vertebrae is maximized, and the serial records based on this orientation constitute a stop-action representation of growth in space as it truly occurs.

Despite these attributes, superimposition on basion has some shortcomings, and it has not been widely adopted. Nevertheless, when viewed from the perspective of basion alignment, enough important features of growth are divulged that an overview of craniofacial growth according to Coben is worth our time.

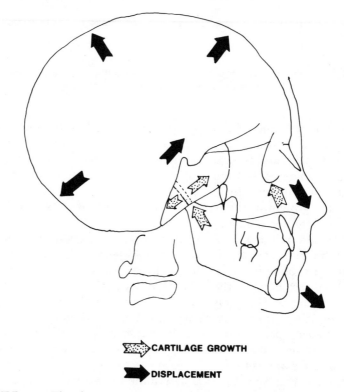

⬚➤CARTILAGE GROWTH

➤DISPLACEMENT

FIGURE 5.2 The directions of displacement growth in the head. Dotted arrows denote contributions made by cartilage replacement mechanisms.

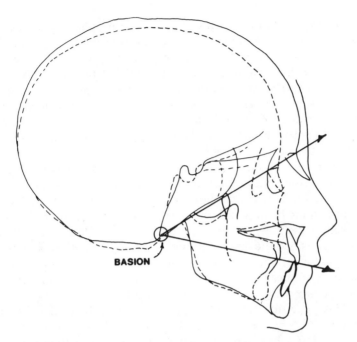

FIGURE 5.3   Serial tracings superimposed on the basion with the anterior
   bases parallel.

Superimposition on the Basion

Coben analyzes serial tracings by registering them on the basion
with the anterior cranial bases parallel. The rationale for using the
anterior cranial base stems from the fact that after the age of seven or
eight it becomes stable and from the thesis (postulated also by
Enlow) that the sight axis of the eye bears a relative constant relationship to the skull.

   With this orientation, craniofacial growth is seen to carry the
dentition spatially away from the vertebral column. The upper face,
housing the maxillary teeth, is carried upward and forward by
growth of the spheno-occipital synchondrosis, whereas mandibular
growth is reflected in a downward and forward movement. Thus,
there are two general vectors of craniofacial growth, and between
these diverging vectors, space is created for vertical facial development (Fig. 5.3).

   From birth to approximately 7 years of age, the distance from
basion to nasion increases as the upper face is carried upward and

forward by a combination of expansion of the spheno-occipital syn-chondrosis and growth that occurs in the anterior cranial base. Both posterior and anterior cranial base growth systems, then, are co-operating at this time to increase the depth of the cranial base and upper face. During this period, the sphenoethmoidal–circummaxil-lary suture system and the nasal cartilage are providing for the forward position of the maxillary complex. Simultaneous mandibu-lar growth approximately equals the combined growth of the cranial base, and the profile comes forward at the same rate.

After the age of seven, there is continued growth of the poste-rior cranial base, but the size of the anterior cranial base (sella to the internal plate of the frontal bone) does not change. Only surface ap-position on the frontonasal region contributes at this time. The max-illa keeps pace with the frontonasal development by surface apposition, since the sphenoethmoidal–circummaxillary suture sys-tem is no longer considered an active growth site. The mandible or-dinarily continues until maturity in a steady downward and forward direction, flattening the profile all the while.

This approach to analysis proposed by Coben, like the other two to follow, emphasizes some things over others. Superimposition on the basion, while deficient as a general technique to evaluate and follow craniofacial growth, does dramatize the role of the cranial base in creating space for facial development.

The Broadbent–Bolton Registration Point

Both cranial base orientations so far discussed, i.e., the S–N plane and Coben's basion method, while revealing a great deal of informa-tion on growth of the head, exaggerate facial growth. The former position thrusts the face downward and forward to the extent that the cranial base grows backward and upward, while the latter dis-places it forward in a divergent pattern to the extent that the poste-rior cranial base grows superiorly and anteriorly. What is needed is the center of growth of the head—that point, perhaps mythical, that can separate cranial from facial growth. The S–N plane registered at the center of sella does not represent this point, nor does the orienta-tion on the basion.

The search for an ideal reference point has proceeded since the inception of cephalometric roentgenology. The classic picture of the growth pattern of the head (Fig. 5.4) is superimposed by the tech-

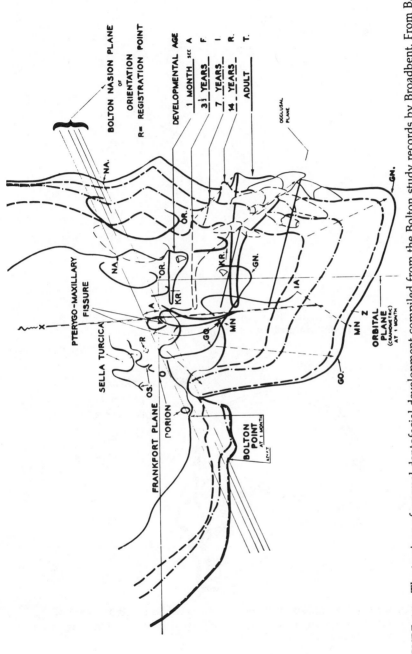

FIGURE 5.4   The tracings of normal dentofacial development compiled from the Bolton study records by Broadbent. From B.H. Broadbent: The face of the normal child. Angle Orthod 7:183, 1937 (with permission).

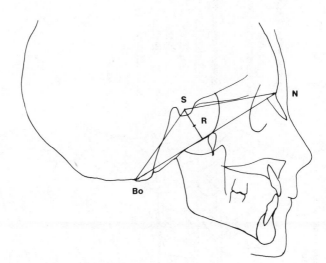

FIGURE 5.5   The registration point within the Broadbent–Bolton triangle used for superimposition of cephalometric tracings.

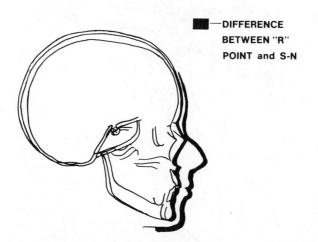

DIFFERENCE
BETWEEN "R"
POINT and S-N

FIGURE 5.6   A comparison of the growth between 6 and 16 years of age when two points of orientation are used. Note that registration on sella exaggerates facial growth.

nique devised by Broadbent for longitudinal studies. This investigator developed an orientation around the cranial base, because it constituted a natural boundary between cranial and facial growth and because it was relatively stable.

The superimposition point on the cranial base is determined by first creating a triangle connecting the sella, nasion and Bolton points (this last point represents the highest point on the profile roentgenogram at the notches on the posterior end of the occipital condyles on the occipital bone). Then, from the sella, a line is dropped perpendicular to the N–BP line (Fig. 5.5). The center of this perpendicular is called the *registration point*, or R. Tracings are superimposed by registering these points and paralleling the N–BP line. The advantage of this process is the separation of sphenoidal growth from facial growth. Figure 5.6 compares the difference in superimposing the tracing from a 6-year-old child and again at 16 years using the two methods of superimposition.

Use of the S–N line exaggerates facial growth compared to use of the registration point. One might argue that this does not really represent a significant difference. However, for research, precise clinical work, and exacting growth prediction, such a difference is detrimental.

Rickett's Pt Point

Through the years there have been a number of reference points and lines proposed to better isolate facial growth from cranial growth. One of the more recent techniques that has gained widespread endorsement is the center of growth and associated planes that have been developed by Ricketts. With modern computer technology in association with serial head-film tracings, he has established that there is a center of least growth (or least change) that can be used as the closest point and orientation for serial reference. This point has been named pterygoid point (Pt) and represents the lower lip of the foramen rotundum as seen in the lateral film. Although this point is probably not the last word in this controversial area, it does permit an informative view of overall craniofacial growth.

To gain a plane for complete orientation a line is drawn from nasion (N) to basion (Ba). The basion is selected as a point near the base of the anterior border of the occipital condyles, at the anterior border of the foramen magnum, or at the end of the clivus plane (Fig. 5.7).

A line perpendicular to Ba–N is drawn through Pt, the intersec-

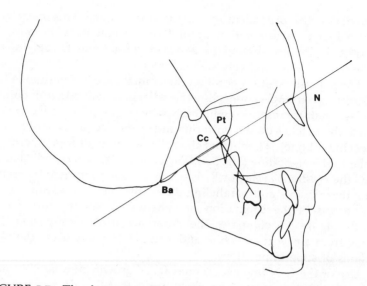

FIGURE 5.7 The derivation of the point Cc (the intersection of Ba–N and the perpendicular through the pterygoid point) used by Ricketts as a point of reference in forecasting the growth of the basion.

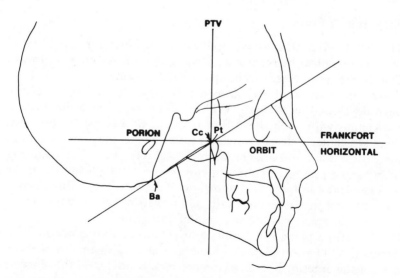

FIGURE 5.8 The intersection of the planes pterygoid vertical (PTV) and the Frankfort horizontal with Ba–N, a point that represents the center of least growth.

tion point being labeled Cc. The forecasting of basion growth away from the center of least growth is made from this point.

If now the Frankfort horizontal (FH) is added to our tracing (a line tangent to the superior edge of the anatomic porion and the inferior border of the orbit) and a perpendicular is erected to this line through point Pt, the picture shown in Fig. 5.8 is obtained. The line perpendicular to the Frankfort horizontal usually lies tangent to the posterior edge of the teardrop-shaped pterygoid fossa and is called the pterygoid vertical (PTV). (Enlow makes good use of a closely related plane, as we will discover later.)

The intersection of FH, PTV, and Ba–N is usually close enough to Pt that, for our purposes, it represents a practical center of least growth; it is that point that, when used as a registration point, depicts maximum growth in every direction.

To get a better perspective of this central point, imagine a tracing of the outline of a child's face on a very thin sheet of transparent rubber. Carefully drawn across the figure is a grid pattern aligned over a duplicate grid in a sheet of paper beneath the piece of rubber (Fig. 5.9). Suppose you drew an adult craniofacial outline on the sheet of paper around the child's outline and stretched the rubber material until the outlines coincided. Upon examination of the squares, it would become quickly apparent that the grid design on the rubber sheet and on the paper no longer matched (Fig. 5.10). Most of the squares on the elastic material would have been displaced and distorted. But there would be one square that moved least. This square would represent the center of least growth for the sheet of rubber. If we wished to establish a universal technique for superimposing stretched rubber sheets, the same square would probably be reasonably close for all cases. Obviously, human craniofacial growth is three dimensional, and considerably more differential than a two-dimensional depiction on a rubber sheet. For this reason, the point of least growth determined by this gimmick would not be expected to coincide with the human point.

An analogy can be made to the study of somatic growth. Referring back to our discussion of the analysis of body height, orientation at the feet limits the information available, because it is somewhat difficult from this perspective to separate the contributions to overall height made by the legs as opposed to the trunk. If the photographs were registered at the sacral region (the approximate center of least growth), the extent of trunk growth superiorly could be easily visualized and isolated from leg growth inferiorly. In effect, this is what a pediatrician does when he measures the sitting height of a child as well as the overall height.

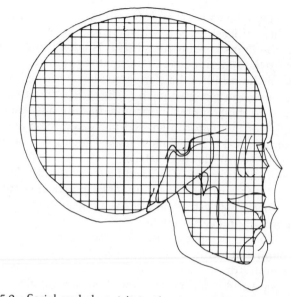

FIGURE 5.9 Serial cephalometric tracings, one on a thin rubber sheet superimposed on the center of least growth. A grid has been drawn on the sheets to demonstrate the distortions that will result when the smaller figure is stretched to match the larger figure.

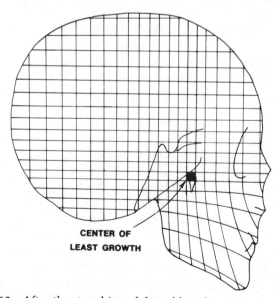

**CENTER OF
LEAST GROWTH**

FIGURE 5.10 After the stretching of the rubber sheet (Fig. 5.9) to make the figures coincide, one square is distorted least. This square represents the center of least growth.

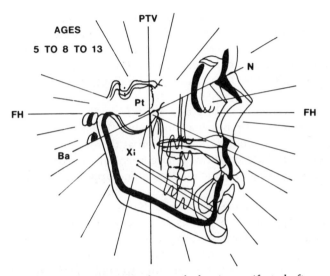

FIGURE 5.11   The "explosion" of growth that is manifested after registration of serial cephalometric tracings according to Ricketts. Adapted from R.M. Ricketts: A principle of arcial growth of the mandible. Angle Orthod 42:368, 1972.

CRANIOFACIAL GROWTH FROM THE CENTER OF LEAST GROWTH

Orientation

By superimposing the serial tracings of an individual according to Ricketts (Fig. 5.11), we can obtain an insight into craniofacial growth. If Pt is indeed the center of least growth (an assumption we will make until further research confirms or refutes it), then there is a wealth of information in longitudinal collections of head film.

One of the first items to note is that the cranial base can be a good reference when the appropriate landmarks are chosen. A plane from Ba to N has two important features: First, it tends to pass over or very near Pt, and second, the anatomic entities labeled Ba and N continue to diverge straight away along the Ba–N plane.

Creating a series of triangles (Ba–S–N) superimposed on Ba–N and registered at Pt (Fig. 5.12) affords an interesting view of cranial base growth. The enlargement of the triangles follows a gnomonic pattern—a type of growth to be discussed in Chapter 7. Briefly, the word describes growth that increases the size of an area or structure without altering its shape.

Notice that the triangles are enlarged superiorly and posteriorly.

If you locate an imaginary vertebra just inferior and posterior to Ba, you can visualize the superior and anterior displacement of the cranial base away from it. This means that although the maxilla is descending vertically from beneath the anterior cranial base, the whole complex is being lifted superiorly by the displacement of the cranial base.

The movement of N up and away from Pt is accomplished by sutural adaption to brain expansion (which ceases at around 8 years of age) and subsequently to apposition at the nasion. The lengthening of Ba–Pt reflects posterior enlargement of the sphenoid, growth at the spheno-occipital synchondrosis, and apposition on the clivus. The relocation of N away from Ba is a result of all these activities, and Pt represents a zone in between that moves least in the process.

Superimposing on S–N tells you how far the maxilla has moved away from the cranial base; orientation on a point like Pt is an attempt to relate how the maxilla and cranial base grow in space as well as to each other. Pt is an attempt to split the difference, so to speak, between all the growths or displacements—superior and inferior, posterior and anterior.

The articulation of the mandible is more closely attuned to posterior cranial base growth than anterior base growth, so that orientation on landmarks associated with the latter (S–N for example) will bias the impressions of growth. Returning to Fig. 5.6, one should be able to visualize how alignment on S–N would exaggerate mandibular protrusion. The problem with such an orientation is that the mandible of the living person is not positioned that way in space. Note that Broadbent's registration point R lies somewhere between S and Pt.

Horizontal Changes

The growth anterior to the PTV plane is generally greater than that posteriorly. This should be expected, since the face is much more immature at birth than the cranium. The mandible, in particular, manifests a large anterior component of growth. The distance the chin point moves from PTV exceeds that of the nasion, thereby establishing a flatter profile. What is also abundantly clear is that the maxilla matches anterior cranial base growth quite closely. The angle Ba–N–A (A represents the most posterior point on the anterior profile of the maxilla above the incisors) remains almost constant throughout the growth process (Fig. 5.13). The anterior component of force generated by the nasal cartilage (or functional matrix) displaces the maxilla in perfect synchrony with brain expansion and nasion apposition.

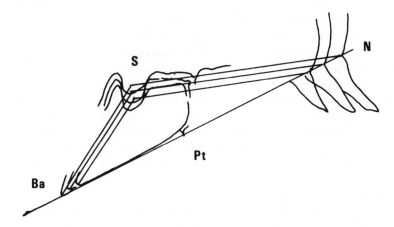

FIGURE 5.12 The growth of the cranial base when serial tracings are oriented at the center of least growth.

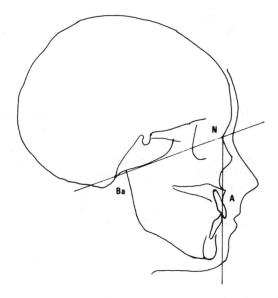

FIGURE 5.13 The angle Ba–N–A, which describes the anteroposterior position of the maxilla relative to the cranial base, which remains virtually constant throughout life.

Brain growth posterior to the PTV plane has some interesting repercussions. The coincidental expansion of the spheno-occipital synchondrosis and its ramifying sutures increases the distance between the glenoid fossae and the center of least growth. As a result, the condyles are also seen to move posteriorly from Pt along the Frankfort horizontal. It must be emphasized that the joint is not actually moving backward; orientation on the center of least growth clarifies that the condylar and ramal growth posterior to PTV is involved in compensating for posterior cranial base growth and not with reducing facial convexity. Thus, part of mandibular growth is concerned with maintaining a relationship and another part with "catching up" to the rest of the face. Incorrect registration of serial tracings fails to separate the two, and consequently, the true change in profile is often distorted.

Compare the progression of tracings of the zygomatic process to those of the orbit and nasion. The greater movement of the latter areas reinforces the previously introduced displacement and reverse drift phenomenon that occurs at the zygomatic process. Anterior resorption prevents the zygomatic process from maintaining its position relative to the orbits and nose, despite its displacement anteriorly with the rest of the maxilla. By adulthood, this process, combined with an increase in the size of the nose, tends to reduce the flatness of the child's face.

Vertical Changes

Turning from the horizontal assessment to the analysis of vertical growth will allow us to examine some other fascinating changes. The Frankfort horizontal, a plane reasonably oriented to the area of least growth, separates superior and inferior growth. According to Scott, the orbit becomes stable at approximately 8 years of age, and at this time, the plane becomes a useful landmark.

The junction between the frontal and nasal bones is carried upward when the inferior border of the orbits are affixed to the Frankfort horizontal. The center of the sella turcica, like the nasion, moves vertically. For this reason, an often-used angle in cephalometrics (S–N–A) would be expected to change insignificantly during a growth.

Because the temporomandibular joint in the glenoid fossa of the temporal bone lies below the expansion that occurs at the spheno-occipital synchondrosis, the distance between the condyles and sella turcica increases. Minor displacement combined with the drift in the middle cranial fossa keeps the condyles close to the Frankfort horizontal throughout growth.

Note the large increase in the anterior vertical dimension of the face (N to Gn). The maxillary complex, while only maintaining a horizontal relationship with the cranial base, exhibits a dramatic vertical movement. Midfacial growth superior to the Frankfort horizontal combined with a significant drop in the palate greatly enlarges the nasal cavity and the capacity for air exchange.

The descent of the mandible from the Frankfort plane is a very orderly process when the corpus axis (Xi–Pm) or "core" of the mandible is analyzed. Because the inferior border of the mandible has shown to be unstable by several investigators, Ricketts has selected two reference points that are more representative of the body of the mandible. These points are Xi, which approximates the mandibular foramen, and Pm, the mental protuberance. The plane connecting these two point represents the stable core of the mandible. The Xi points of serial tracings lie on the same polar coordinate that radiates from Pt; the Pm points follow yet another coordinate. The corpus axis expands between these diverging coordinates as it drops vertically, remaining virtually parallel in the process.

Because, according to Moss, the volume of the oropharyngeal cavity is a genetically determined requisite for life and not just that space which happens to be left over after soft-tissue formation, the mandible must be able to accelerate its growth at appropriate times. The volume of the oral cavity and pharynx must enlarge with somatic growth to handle the demands for air and nutrients, and as a consequence, the mandible must respond in kind. Concurrently, it must keep pace with the vertical and horizontal growth of the maxilla and the eruption of teeth in their alveolar processes. Considering all this, it is a wonder that we do not see more anteroposterior malocclusions of the jaws than we do.

The pattern of overall facial growth from the perspective just described suggests that little is left to chance. The design for growth along certain planes (Ba–N and FH) indicates an orderliness that might not be apparent from casual observation. Although there is tremendous growth, the face of a man looks almost the same, regardless of age. This growth is gnomonic—a fact of importance in growth prediction.

The maxilla and mandible grow under the cranial base, which is increasing its horizontal dimension in response to brain expansion. The maxilla, related to the anterior cranial base, must synchronize its growth horizontally and vertically. The mandible, articulated in a sense to one end of the cranial base (condyles in the glenoid fossa), functions with respect to the maxilla—an anterior cranial base-related complex. Thus, the mandible must grow enough to compensate for posterior cranial and anterior cranial base growth, but also enough to flatten the profile and increase the height of the face.

The orientation of head film tracings on polar coordinates

allows us to visualize all these concurrent happenings as an "explosion" from a center of least growth. Not only is this technique visually very informative, but it has furthered the science of cephalometrics and growth prediction.

SUMMARY

In Table 5.1 are listed the essentials of craniofacial growth. The head is divided in three regions—neurocranium, middle cranial fossa, and lower face—separated here into horizontal boxes. Vertically, these divisions are subdivided into counterparts. Anteriorly, the counterparts consist of the anterior cranial fossa, anterior cranial base, ethmomaxillary complex, and corpus. Posteriorly, they are the middle cranial fossa, posterior cranial base, oropharyngeal cavity, and ramus.

Most of the mechanisms of bone growth are used in each anatomic division with the exception of sutures. While the cranium and middle face manifest a number of these sutures, none exists in the mandible after birth.

Each of the regions embodies a type of cartilage replacement mechanism. The posterior cranial base houses the spheno-occipital synchondrosis, a bidirectional bone-forming structure. In the middle face, the nasal cartilage is thought by many to thrust the maxilla downward. And in the lower face, the mandible is aided in its growth by a unique cartilage growth mechanism, the condyle.

Apposition and resorption are processes used universally in the head. Each anatomic division can also be analyzed for examples of the principles of bone growth. Drift versus displacement situations are replete in all areas. For instance, in the cranium, the bones of the cranial vaults are displaced centrifugally by brain growth; external apposition abets this process. In the middle face, the palate moves inferiorly by displacement of the whole complex, by resorption on the nasal surface, and by apposition on the oral side. Prominence of the symphysis of the mandible is a result of both processes.

The phenomenon of growth in one direction and movement in another is manifested in each area. The spheno-occipital synchondrosis contributes to inferior growth of the occipital bone, with a superior growth of the cranial base resulting. Posterior growth–anterior displacements describe the growth of the maxilla in the middle face and the mandible in the lower face.

The V principle finds representation in several aspects of maxillary growth and the ramal and condylar development of the mandible.

TABLE 5-1   The Essentials of Craniofacial Growth

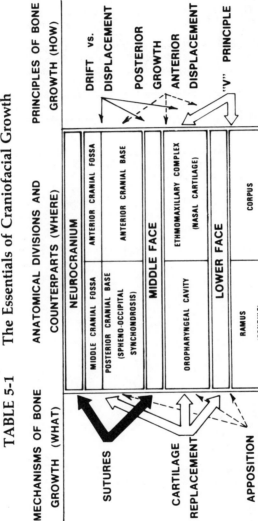

BIBLIOGRAPHY

Baume LJ: A biologist looks at the sella point. Report of the Thirty-third Congress, Trans Eur Orthod Soc. p 150, 1957

Bjork A: Cranial base development. Am J Orthod 41:198, 1937

Broadbent BH: The face of the normal child. Angle Orthod 7:183, 1937

Cannon J: Craniofacial height and depth increments in normal children. Angle Orthod 40:202, 1970

Coben SE: Growth and class II treatment. Am J Orthod 52:5, 1966

Enlow DH: Handbook of Facial Growth. Philadelphia, Saunders, 1975

Latham RA: The sella point and postnatal growth of the human cranial base. Am J Orthod 61:156, 1972

Ricketts RM: Planning treatment on the basis of the facial pattern and an estimate of its growth. Angle Orthod 27:14, 1957

Ricketts RM: Cephalometric synthesis. Am J Orthod 46:647, 1960

Ricketts RM: A principle of arcial growth of the mandible. Angle Orthod 42:368, 1972

Ricketts RM: New perspectives on orientation and their benefits to clinical orthodontics—Part I. Angle Orthod 45:238, 1975

Ricketts RM: New perspectives on orientation and their benefits to clinical orthodontics—Part II. Angle Orthod 46:26, 1976

Scott JH: The growth of the human face. Proc R Soc Med 47:01, 1954

· 6 ·

Theories of
Craniofacial Growth

We dance round in a ring and suppose
But the secret sits in the middle and knows.
Robert Frost, The Secret Sits

In Table 5–1, the essentials of craniofacial growth, the where, what, and how are listed. The divisions of the head help explain where growth is occurring; the mechanisms of bone growth relate what tissues are contributing; the principles of bone growth describe how the growth is manifested.

So far only alluded to are the why questions. The all-encompassing, mind-boggling question of craniofacial biology is, "Why does the head grow as it does?" Subordinate to this major question are all the other whys: "Why do humans look like humans?" "Why do sutures respond as they do?" "Why does the face grow downward and forward?" "Why is an eyeball essential for orbit development?" "Why does the concave bending of bone elicit bone deposition?" "Why is the presence of teeth required for the existence of alveolar bone?" "Why do some pluripotential cells become osteoblasts and others osteoclasts?" The questions could go on and on.

Factual data in this realm of inquiry is minimal; theories are beginning to emerge, but they are operating mainly at the macromolecular level. The ultimate role of genetic information in the determination of why the face grows as it does has only been superficially studied.

GROWTH SITES VERSUS GROWTH CENTERS

Definitions

Let us begin to explore where craniofacial biologists have taken us in this fascinating field. Consider the distances from points A to B across the growth plate and C to D across a suture (Fig. 6.1) as marked by experimental metallic implants.

In situ, both distances are capable of increasing sometime during growth; we can find many examples of each. But why do the markers separate? From previous discussion, we know that the cartilage cells of the growth plate are capable of creating a tissue-separating force by virtue of interstitial expansion. When such units are transplanted to subcutaneous sites, the dimension A–B continues to increase.

The cells of the suture, however, are not capable of generating a tissue-separating force; they fill in the void as the bone segments are pulled apart. When transplanted to a subcutaneous site, the implants C and D no longer grow apart.

Baume has given names to these two growth mechanisms. "Places of endochondral ossification with tissue separation force,

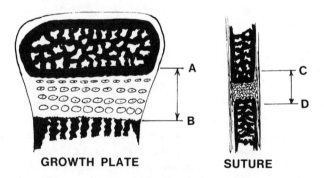

GROWTH PLATE   SUTURE

FIGURE 6.1 The growth plate is an example of a growth center; the suture has been named a growth site. The distance between A and B and between C and D increase for different reasons.

contributing to the increase of skeletal mass" he has called growth centers. "A region of periosteal or sutural bone formation and modeling resorption adaptive to environmental influences" is a growth site.

It might reasonably follow that the extension of this classification to the growth mechanisms of the head would be a clarifying move. This has not been the case. While most authorities would agree that the sutures are growth sites, the various cartilage replacement mechanism are not universally accepted as growth centers. In fact, one theory maintains that all of them are growth sites.

The Suture as a Growth Site

To help explain the origin of this ongoing controversy, a little historical perspective might be in order. Not too many years ago, two concepts dominated the thoughts in this area, mostly by default. It was assumed that the sizes and shapes of bones were genetically determined. As a corollary to this general principle, sutural patterns were necessarily predetermined. In the head, where bone size and shape is profoundly influenced by sutures, it must follow that sutural positions limit and control expansion of bones.

Moss took these assumptions to task by designing an experiment to test the predetermination of the cranial sutures. While infantile, the sutures of the heads of rat pups were damaged by fine instruments. Moss's thesis was that if the pattern of the sutures were genetically determined and inviolate, any death of cells in the structure would leave a defect in the hard-tissue covering. On the other hand, if the information determining the size and shape of a bone does not reside within the cells of bone or the sutures, then the drive to keep a protective covering over the brain should force compensatory growth of adjacent bones. In Fig. 6.2A, the damaged suture does not result in a void; instead, the undamaged bone crosses the midline (Fig. 6.2B) to compensate.

This experiment clearly demonstrated that sutures are very flexible. How, then, can there be sufficient genetic information to tell the adjacent bone where to compensate? Obviously there cannot be. The variety of possible experimental insults would rapidly outstrip available programs within the chromosomes. Clearly, some other forces are determining the growth of bones, and the sutures are merely tools used in this process.

This concept does not preclude genetic information from determining the size and shape of bones; it only states that the information resides in cells other than those of bone.

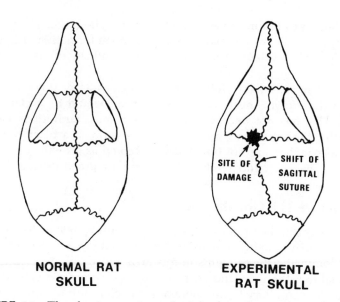

**NORMAL RAT
SKULL**

**EXPERIMENTAL
RAT SKULL**

FIGURE 6.2   The classic experiment by Moss entailing cautery of the cranial sutures of rat pups, which disproved the predetermination theory of sutural positions.

The Condyle as a Growth Site—A Biological Basis?

The one cartilage replacement mechanism that seemingly qualifies as a growth site is the condyle of the mandible. This designation means that these growth areas do not push the rest of the mandible downward and forward, but instead "fill in" (as do the sutures) when some other influence displaces it. Studies on primates have verified the adaptability of the cells of the condyles. Pressure relief at the temporomandibular joint engendered by appliances that held the mandible forward elicited a burst of cell division and formation of new matrix. It is not unreasonable to speculate that natural forces could similarly displace the mandible during growth and maintain a level of condylar activity.

The potential for a condyle to act differently than a growth plate has, at least, a biologic basis. From the writings of Baume, Durkin et al., and Brigham et al., a list of the differences between condylar and epiphyseal cartilages has been compiled (Table 6.1).

Some of the biologic criteria have already been discussed, but others could probably use further amplification. The origin and growth of the two sites have been covered. With respect to maturation, Durkin has classified the cartilage of the condyle as embryonic like the anlage of a long bone, rather than specialized like the growth

TABLE 6-1* Principles of Cepholafacial Development

BIOLOGICAL CRITERIA	EPIPHYSEAL GROWTH PLATES	CONDYLES
Origin	Derivative of primordial cartilage	Secondary cartilage formation on original membrane bone
Growth	Interstitial, three dimensional in hyaline cartilage	Peripheral in fibrocartilage covering; proliferating cells are not cartilage cells but are rather like undifferentiated mesenchymal cells
Maturation	Secondary ossification centers, final fusion, disappearance of all cartilage	Conversion from hypertrophic to nonhypertrophic state—but no complete conversion to bone
Histology	Only the degenerative zone is mineralizing; primary spongiosa	Whole hypertrophic area in state of mineralization; no primary spongiosa; structural organization lacking
Hormonal control	Marked response to thyroxine deficiency; after final fusion no further response to growth hormone	Minimal response to thyroxine deficiency; matured condyle can be reawakened by growth hormone
Vitamin response	Ascorbic acid deficiency leads to gerüstmark zone; vitamin D deficiency results in classic picture of rickets	Ascorbic acid deficiency elicits minimal response; vitamin D deficiency causes reversion to more immature stage
Mechanical stimuli	Unresponsive	Responsive
Antigenic differences	Possesses antigenic determinants common to condylar cartilage and nasal septum	Possesses one or more unique antigenic determinants distinct from the epiphyseal cartilages and nasal septum

* Table modified from L. J. Baume: Principles of cephalofacial development revealed by experimental biology. Am J Orthod 47:881, 1961.

plate. The latter tissue, postpubertally, transforms completely to bone and causes linear growth of an individual to cease.

The condyle is hypertrophic or immature throughout its growth period. The cellular organization is haphazard, matrix formation is scanty, the calcification is pericellular, and the trabecular formation does not resemble the ordinary endochondral growth process. On maturation a nonhypertrophic stage is reached where cell activity diminishes, cartilage erosion ceases, and the bones apposes the cartilage abruptly, forming a kind of bony seal. The cartilage is not, however, replaced by bone.

It can be stated summarily that the growth plates are decidely more sensitive to hormone and vitamin deficiency than the condyles. In contrast, the latter structures are far more responsive to mechanical stimuli; pressure diminishes cell activity, while relief from pressure promotes it. The growth plates, on the other hand, can expand under considerable load, as represented by either physiologic weight or orthopedic devices, which span the cartilage in an attempt to restrain growth.

The studies exploring the antigenic potential of the various cartilages has disclosed that condylar cartilage contains a unique proteoglycan. This biochemical difference (and others like it) might someday help explain the more obvious functional and morphologic differences.

While there might remain some voices claiming that the condyles are growth centers, the bulk of experimental and clinical evidence suggests strongly that these structures respond as growth sites most of the time.

THE FUNCTIONAL MATRIX

Definition

Based on the original concepts of van der Klaauw, his own experimental work and that of others, combined with clinical interpretations and experiences, Moss has formulated the functional matrix theory. Succinctly stated, the theory is as follows:

> There is no direct genetic influence on the size, shape, or position or skeletal tissues, only the initiation of ossification. All genetic skeletogenic activity is primarily upon the embryonic functional matrices.

Let us examine this profound statement piece by piece. The first part, the lack of direct genetic influence, has just been introduced in the discussion of sutural adaptability. Moss has extended the concept to the fullest; the only manifestation of intrinsic genetic direction by bone cells is the process of ossification. The orientation and spatial relationship of bone cells and their origin from precursor cells are all determined by outside influences. The environment selects and directs the cells concerned with bone formation. Only when the circumstances are appropriate for bone formation does the DNA of these bone cells set in motion those activities involved in ossification. Thus, in the ontogeny of the individual, there is no set number of cells earmarked to differentiate into osteoblasts and osteoclasts. Rather, other tissues and functions provide the impetus for differentiation at the right time and in the right places. These tissues and functions have been called the embryonic functional matrices. This concept is not in every instance easy to understand.

Functional Components and Skeletal Units

According to Moss, the head is a structure designed to carry out functions:—neural integration, respiration, digestion, hearing, olfaction, and speech—to name the important ones. Each of these functions is accomplished by certain tissues (and spaces) in the head. The tissues and spaces responsible for a single function have been termed a functional cranial component. Thus, the component handling speech would consist of the lips, teeth, tongue, oral cavity, nasal cavity, etc.—any aspect of the head that enables a person to speak is part of this functional component. Moss divides these components into two parts—a functional matrix consisting of the soft tissue and space that completely performs a particular function, and a related skeletal unit that acts biomechanically to protect and/or support its functional matrix.

Referring again to the component responsible for speech, the maxilla and mandible would constitute the skeletal unit supporting the matrix. The implication of this relationship is that the skeletal unit is subordinate to and supportive of the functional matrix; the bone tissue assumes a size and shape that best enables the matrix to perform its function. As a consequence, the cells of bone need not have genetic information for morphologic orientation; the functional matrix will provide the direction.

To better understand how the matrix can influence the form (size and shape) of bone, it is advantageous to think in terms of two types of functional matrices and two types of skeletal units (Fig. 6.3).

The first type of functional matrix has been named periosteal. This term relates the matrix to those tissues that influence the bone directly through the periosteum. Muscles are attached to the periosteum and consequently are excellent examples of this kind of matrix.

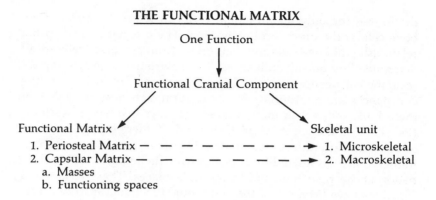

FIGURE 6.3   Classification of the functional matrix.

Blood vessels and nerves lying in grooves or entering or exiting through foramen can also exert a "periosteal" influence on the skeletal unit.

The periosteal matrix affects a microskeletal unit, meaning that the sphere of influence is usually limited to a part of one bone. The temporalis muscle exerts most of its action on the coronoid process, a microunit of the mandible. A tooth is responsible for the alveolar bone that supports it; extract the tooth (the periosteal functional matrix) and the microskeletal unit (the immediate alveolar process) disappears.

The second type of functional matrix is termed capsular. Certain aspects of this type of matrix will probably fit well into our concepts of what makes things grow; other aspects are more difficult to visualize. Included in this class of matrix are those masses and spaces that are surrounded by capsules. For instance, the neural mass is contained within a capsule of scalp, dura mater, etc., and the orbital mass is surrounded by the supporting tissues of the eye. The oral–nasal–pharyngeal spaces are surrounded by a variety of tissues that compose their capsules.

These capsules tend to influence macroskeletal units, which means that portions of several bones are simultaneously affected. An excellent example is the inner surface of the calvarium. We have already discussed the apposition–resorption pattern of the cranial vault, and it should be apparent that it has nothing to do with a specific bone. The neural mass within its capsules elicits a reaction on the surfaces of the calvarium that transcends a localized area. As a result, apposition on the occipital, parietal, temporalis, and frontal occurs as if all of them were but one bone. This sharing of reaction by several adjacent bones constitutes a macroskeletal unit.

The Functional Matrix by Regions

With these definitions the growth of the head can be analyzed from a new perspective. The functional matrix theory is probably the dominant theory in craniofacial biology, even though every aspect of it is not universally accepted. However, using it as a frame of reference facilitates the introduction of other theories, since most of them are exceptions to it in one or more ways.

A strong defense for the existence of periosteal functional matrices can be made. After all, the reaction in the related microskeletal units is transformation, which can also be called drift, apposition–resorption changes, or intraosseous remodeling. Myriads of bone transformations occur as muscles, nerves, glands, soft tissues, etc. enlarge and change positions during growth of the head.

It is also not too difficult to envisage the role of the brain in the expansion of the neural capsule. As the neural mass enlarges, the bones within the capsule are displaced outward in a process termed translation. The sutures are not separating the bones by the pressure of cell proliferation; instead, they are preventing the creation of voids by filling in the separating areas. Thus, a functional matrix influences a macroskeletal unit, the entire cranium.

Where the role of functional matrix becomes most difficult to visualize is in development of the face. According to Moss, the nasal cartilage and the condyles of the mandible are growth sites and therefore incapable of tissue-separating force. As a consequence, the translation of the middle and lower face downward and forward must be accomplished by the oral–nasal–pharyngeal capsules. The soft tissues of these capsules are of necessity the determinant of their size and position in space. The skeletal units only respond, offering continually adapting biomechanical support. The factor that dictates the size of the facial capsules is the volume of the functioning spaces. This concept is a reversal of our usual thinking. It probably has never occurred to us that a space might have priority over the tissues that surround it. Bear in mind, however, that space is not something that happens to be left over after everything else has formed. The patency and adequacy of oronasal tubes are so fundamental that nature programs their size and guarantees that the increased demands of somatic growth are met by craniofacial expansion. If you doubt, for example, the relationship between the functioning spaces of the face and the requirements for air exchange, simply close one nostril with finger pressure and proceed to breathe. So uncomfortable does this simple act become that panic can begin to set in.

Moss contends, then, that all the loci of new bone formation (sutures, periosteum, spheno-occipital synchondrosis, nasal cartilage, and condyles) are growth sites and not growth centers. None of these sites contains the genetic information that can determine their ultimate form; all of them are at the disposal of the functional matrices related to them. Moss has removed the cartilage replacement mechanisms out of the category of Baume's growth centers, and not every one agrees with this.

The dependence of cranial vault bones on the neural mass beneath them in reaching their final form can be demonstrated by two extremes of clinical experience. In the condition of microcephaly where the fetus develops an inadequate brain, the size of the cranial vault is correspondingly small. At the other end of the spectrum, an enlarged neural mass, which can result from meningeal fluid retention (hydrocephaly), effects a calvarium of similar proportions.

Not so responsive to these changes in the neural functional matrix is the spheno-occipital synchondrosis. The size attained by the posterior cranial base is not so closely attuned to the tissues it supports. Additionally, subcutaneous transplants of animal synchondroses have demonstrated some degree of autonomous growth. Thus, there appears to be some measure of tissue-separating force within this structure, and what may ultimately be discovered is that it is a modifiable growth center.

Another cartilage replacement mechanism, thought by some to be a growth center, is the nasal cartilage. Scott, in particular, conceived of midfacial growth in response to the interstitial expansion of this structure. This concept refutes the functional matrix theory, and the controversy has been joined.

Numerous experiments to ascertain the true role of the nasal cartilage have been performed, but the studies and the interpretations have been conflicting. There is no doubt that the surgical removal of the spectum predisposes the experimental animal to a shortened nostril and a prognathism. This result would strongly suggest a dynamic role of the nasal septum in nostril development, but others have concluded that such intervention succeeds only in creating enough scar tissue to prevent the area following the flow of its functional matrix.

One experiment designed to destroy cell proliferation potential while avoiding cicatrization has claimed to show development of the midface without the tissue-separating force of the septum. Unfortunately the differences have not been resolved, and one investigator has suggested that for the time being the nasal cartilage can be thought of as the symbol of midfacial growth.

The remaining cartilage growth loci in the head are the condyles of the mandible. According to the functional matrix theory, the orofacial capsule translates the mandible away from its articulation in the glenoid fossa. The lack of pressure on the articular surface presumably stimulates the generation of new cells and matrix sufficient to maintain a good fulcrum. When the person has reached full size, the capsule ceases to enlarge, the mandible is no longer translated, pressure increases on the articular surfaces, and the mitotic activities slacken.

Moss offers as evidence for this sequence of events the type of growth manifested by patients from whom the condyles have been removed, which is sometimes done when trauma or pathology has ankylosed the temperomandibular joint. Despite the loss of the growth activity of the condyle, the mandible achieves a normal size and translates downward and forward within its capsule. Although muscular adaption to the lack of fulcrum points modifies the shape somewhat, the ramus and corpus apparently grow and remodel without restraint.

Experimental studies that consisted of cast-gold bite planes fitted to the upper jaws of primates and fashioned in such a way as to hold the lower jaw in protrusion have suggested that the cartilage, when relieved of pressure, undergoes a burst of activity. The condyles, then, tend to grow back to the fossa to restore contact. Either natural translation (the functional matrix) or experimental displacement (bite plane) elicits growth activities, both characteristic of a growth site. Apparently, the cells of this cartilage replacement mechanism have the genetic information that enables them to be responsive to pressure; they do not, however, have the capacity to determine their own size. Their functional matrix decides this.

Current Clinical Applications of the Functional Matrix

Lest this discussion on the theories of craniofacial growth become too heady for the clinically oriented, it would probably be worthwhile to note the present widespread utilization of the functional matrix, albeit disguised by other names. Consider the correction of malocclusion by intraoral and/or extraoral appliances (Fig. 6.4).

ORTHODONTIC THERAPY INVOLVES A CHANGE IN:

1. Periosteal Matrix ⟶ Skeletal Unit
   (Teeth)   (Alveolar Bone)

and/or

2. Capsular Matrix ⟶ Several Skeletal Units
   (Orofacial Orthopedics)   (The Jaws)

FIGURE 6.4 Use of the functional matrix in routine orthodontics.

Whether realized or not, forces exerted on teeth or jaws do influence their functional matrices; and there are still other matrices that determine the stability of the corrected malocclusions. To modify one functional matrix at the expense of another invites trouble. Relapse or perpetual retention is often the choice following orthodontic therapy.

The following is a list of current clinical practices that reflect some aspect of functional matrix modification.

1. Enucleated orbit. The replacement of eyes with protheses that are periodically replaced by larger versions promotes growth of the orbit.
2. Widening midpalatal sutures. The rapid palatal expansion (RPE) appliance is much in vogue these days to widen the maxillary arch in cross-bite malocclusions. This therapy is a form of orofacial orthopedics.
3. Repositioning of the maxillary segments of cleft patients. These alignment procedures, like the preceding example, involve the change of macrounits.
4. Bilateral condylectomy. When ankylosis of the condyles occurs in the growing child, condylectomy removes the restraints and allows the maximum development of the mandible in space.
5. Oblique bite plane. Intraoral devices that hold the mandible in a protruded position for the purpose of stimulating condylar growth.
6. Monobloc functional therapy. Intraoral appliance used in conjunction with bone grafting to stimulate mandibular bone remodeling following subtotal mandibular resection.

VAN LIMBORG'S VIEW OF CRANIOFACIAL GROWTH

From the diversity of opinion about the autonomy of certain bone-forming loci, van Limborg has distilled the essential theories, tested them in the light of experimental and clinical evidence, and forged a compromise theory of his own.

He has categorized three major viewpoints and ascribed each of them to one principal spokesman. Thus, the theories advanced are those of Sicher, Scott, and Moss, but we should be aware that the work of countless others has helped formulate each of them.

Before a step-by-step analysis of each theory, it would be beneficial to become acquainted with the anatomic division used by van Limborg and also to define the factors that he avers control growth.

According to Van Limborg, the embryologic origin of the components of skull determines the kind of growth that occurs there. The cartilaginous base, the nasal capsule, and the otic capsules are sites of endochondral ossification and become known as the chondrocranium. All the bones rising from these cartilage precursors have, for varying degrees of time, the capability for interstitial expansion while they are growing.

Direct deposition of bone, intramembranous ossification, forms the calvarium, middle face, and mandible—an aggregation called the desmocranium.

In Table 6.2 are listed the factors judged by van Limborg to control craniofacial growth. Moyers' listing of definitions has embellished the original paper. (Information added by the present author has been set in italics in order to relate the new terms of van Limborg to previous concepts.)

TABLE 6.2*   Controlling Factors in Craniofacial Growth

DEFINITION	FACTOR
Intrinsic genetic factors	Genetic factors inherent to the skull tissues
Local epigenetic factors (*Capsular functional matrix*)	Genetically determined influence originating from adjacent structures *and spaces* (brain, eyes, etc.)
General epigenetic factors	Genetically determined influences originating from distant structures (sex hormones, etc.)
Local environmental factors (*Periosteal functional matrix*)	Local nongenetic influences originating from the external environment (local external pressure, muscle forces, etc.)
General environmental factors	General nongenetic influences originating from the external environment (food and oxygen supply, etc.)

* Modified from Moyers RE: Handbook of Orthodontics, 3rd ed., Chicago, Year Book, 1973.

Sicher

Sicher represents the older concept of craniofacial growth that holds that the destiny of skull tissue is controlled largely by its own intrinsic genetic information. The only influence demonstrable as a result of external factors is localized external and internal remodeling. Diagrammatically, this classic view of skull growth is depicted in Fig. 6.5.

In Sicher's view all bone-forming elements (cartilage, sutures, and periosteum) are growth centers. For example, the sutures attaching the maxillary complex to the cranium can both drive the

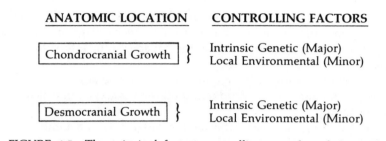

FIGURE 6.5 The principal factors controlling growth and their relative impact on the two embryologic divisions of the head according to Sicher.

midface down by virtue of cellular proliferation and also determine the extent of this activity through their own genetic composition.

This classic theory fails because the independence of the skull growth cannot be consistently demonstrated. Consider again the development of the calvarium in cases of microcephaly and hydrocephaly. The ability of the neurocranium to just cover the brain (and no more) is not in accordance with a structure whose size is genetically determined.

Another organ that has a similar influence on the bone around it is the eye. Experimental studies have demonstrated in a variety of ways that the orbits exist only to house the eye. Manipulation of the primordia of the eye in the embryo can create an animal with one orbit, two orbits either very close or abnormally far apart, or even an animal with three orbits. Postnatally, the eye continues to profoundly affect the surrounding bone. If the eye is enucleated and not replaced by a prosthesis (enlarged periodically during growth), the orbit will cease to expand.

These and other data are not consistent with the tenets of an all-intrinsic genetic theory.

Scott

Scott has already been mentioned with regard to the nasal cartilage. His full theory expounds cartilage and periosteum as growth centers and classifies sutures as passive and secondary. Van Limborg finds fault only with Scott's implication that periosteal tissue is dominated by intrinsic genetic factors. In Fig. 6.6 is diagrammed the essence of Scott's theory.

Scott has correlated sutural growth with synchondrosis activity and with other tissue growth such as the brain. Some local factors could also possibly modify the process. Van Limborg feels that the periosteum should fall into the same category as the sutures, since

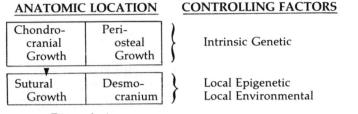

ANATOMIC LOCATION   CONTROLLING FACTORS

| Chondro-cranial Growth | Peri-osteal Growth | } | Intrinsic Genetic |
| Sutural Growth | Desmo-cranium | } | Local Epigenetic Local Environmental |

⟶ Direct Action

FIGURE 6.6   The principal factors controlling growth that operate on the divisions of the head according to Scott.

there cannot reasonably be that much difference in their cellular profile.

Moss

The functional matrix theory is challenged by van Limborg for several reasons. Moss's placement of cartilage replacement mechanisms in the same category as sutures and periosteum is not supported by clinical evidence. Conditions of microcephaly, anencephaly, and hydrocephaly, which have great impact on the desmocranium, do not significantly change the cranial base. Evidence for this statement can be seen in Table 6.3, which has been compiled from several sources. On the other hand, the condition of premature fusion of the calvarial sutures, craniostenosis, suggests to van Limborg that sutures have some capacity to regulate their activity. Thus, Moss is felt to have erred in denying any intrinsic genetic factors in the control of chrondrocranial growth and restricting the control of sutural growth

TABLE 6.3   Dimensions of Two Microcephalic and Two Hydrocephalic Skulls Compared to Normal Skulls*

	MICRO-CEPHALIC		HYDRO-CEPHALIC		RANGE OF 300 NORMAL MALE SKULLS
	1	2	3	4	
Cranial capacity (ml)	455	695	2660	2980	1,170–1,930
Cranial length (mm)	140	153	218	214	171–204
Cranial width (mm)	103	108	176	181	128–152
Cranial base (mm)	94	92	98	112	86–112

* From Scott: The growth of the human face. Proc R Soc Med 47:91, 1954.

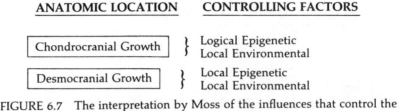

FIGURE 6.7   The interpretation by Moss of the influences that control the growth of the head.

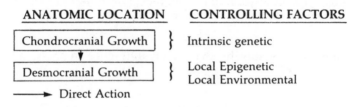

FIGURE 6.8   The compromise theory of van Limborg that was formulated to better fit the clinical and experimental data.

to local epigenetic and environmental factors. Figure 6.7 is the diagrammatic representation of Moss's theory.

Van Limborg's Compromise

After a survey of the preceding theories, van Limborg has summarized the following as the essentials of craniofacial growth.

1. Chondrocranial growth is controlled by intrinsic genetic factors.
2. The desmocranial growth is controlled mainly by many epigenetic factors that originate from skull cartilages and other head tissues.
3. The growth of the desmocranium is influenced by local environmental factors occurring in the form of tension forces and pressures.
4. General epigenetic and general environmental factors are of minor importance. Van Limborg's controlling system can be diagramatically represented as in Fig. 6.8.

A Modern Composite

Because van Limborg used a few terms to describe embryologic entities that differ from those previously introduced in this book, and because he fails to classify the controlling factors for the mandible, the composite of craniofacial growth in Fig. 6.9 is offered.

Although the calvarium and the face arise from membrane

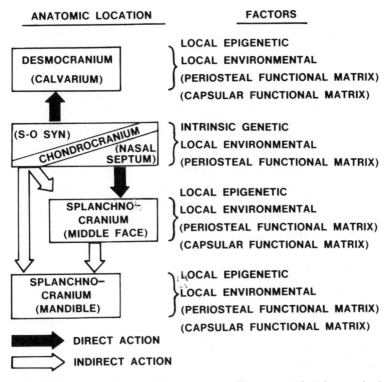

FIGURE 6.9 A composite of factors controlling craniofacial growth that seem to agree with current concepts and data.

bone, most embryologists separate them on the basis of embryologic origin. The desmocranium (or neurocranium) develops directly around the brain; the middle face and mandible are derivatives of the first branchial arch in a process involving complex migration and fusion of processes. These components are often listed as the splanchoncranium.

As diagrammed, the chondrocranium is the dominant factor in craniofacial growth. The postnatal cartilage remnants, the spheno-occipital synchondrosis, and the nasal cartilage act as growth centers, largely influenced by intrinsic genetic factors. The spheno-occipital synchondrosis exerts a direct action on the desmocranium, which is also dominated by brain expansion. Thus, the calvarium is controlled by local epigenetic and environmental factors. The sutures are growth sites, and the entire process of cranial vault growth substantiates the concept of the functional matrix.

The other cartilage mechanism of the original chondrocranium, the nasal cartilage, displaces the maxilla downward and forward.

The possibility of a capsular functional matrix exercising this overall control is open to question. However, the orbits are regions of the middle face that are undeniable subjects of a functional matrix.

The mandible appears to be dominated by local epigenetic and environmental factors, and consequently the concept of Moss fits comfortably into this area of development. The white arrow from the cranial base to the mandible denotes that there is an indirect relationship between the two. The cranial base flexure and growth alters the position of the glenoid fossa and thereby the spatial position of the mandible. This is distinctly different from the effect of the nasal cartilage on the maxilla. The vertical position of the maxilla with respect to the glenoid fossa as determined by the flatness of the cranial base can cause rotation of the mandible forward or backward. Thus, the mandible is influenced—but not controlled—by those bones with which it articulates.

BIBLIOGRAPHY

Baume LJ: Principles of cephalofacial development revealed by experimental biology. Am J Orthod 47:881,1961

Baume LJ, Derichsweiler H: Is the condylar growth center responsive to orthodontic treatment? Oral Surg 14:347, 1961

Berkman MD, Rothschild D, Trieger N, Goldman A: Clinical application of the functional matrix: mandibular reconstruction and monoblock functional therapy. Am Dent Assoc 96:645, 1978

Brigham GP, Scaletta LJ, Johnston LE Jr, Occhino JC: Antigenic differences among condylar, epiphyseal, and nasal septal cartilages. In McNamara JA Jr (ed): The Biology of Occlusal Development. Ann Arbor, University of Michigan, 1977

Dorenbos J: In vivo cerebral implantation of the anterior and posterior halves of the spheno-occipital synchondrosis in rats. Arch Oral Biol 17:1067, 1972

Durkin JF, Heeley JD, Irving JT: The cartilage of the mandibular condyle. Oral Sci Rev 2:29, 1973

Koski K: Cranial growth centers. Facts or fallacies? Am J Orthod 54:566, 1968

Koski K: The role of the craniofacial cartilages in the postnatal growth of the craniofacial skeleton. In Dahlberg AA, Graber TM (eds): Orofacial Growth and Development. Hague, Mouton, 1977

Koski K, Ronning O: Growth potential of intracerebrally transplanted cranial base synchondroses of the rat. Arch Oral Biol 15:1107, 1970

Koski K, Ronning O: Growth potential of subcutaneously transplanted cranial base synchondroses of the rat. Acta Odontol Scan 27:343, 1969

Klaauw CJ van der: Size and position of the functional components of the skull. A contribution to the knowledge of the architecture of the skull, based on data in the literature. Arch Neerl Zool 9:176, 1948

Kremenak CR, Hartshorn, DF, Demjen SE: The role of the cartilagenous nasal septum in maxillofacial growth: Experimental septum removal in beagle pups. J Dent Res 48:abstract 32, 1969

Limborg J van: A new view on the control of the morphogenesis of the skull. Acta Morphol Neerl Scand 8:143, 1970

Meikle MC: The role of the condyle in the postnatal growth of the mandible. Am J Orthod 64:50, 1973

Moss ML: Growth of the calvaria in the rat. The determination of osseous morphology. Am J Anat 94:333, 1954

Moss ML: The primacy of functional matrices on orofacial growth. Dent Pract 19:65, 1968

Moss ML: Twenty years of functional cranial analysis. Am J Orthod 61:479, 1972

Moss ML, Greenberg SN: Functional cranial analysis of the human maxillary bone. Angle Orthod 37:151, 1967

Moss ML, Bromberg BE, Chul Song I, Eisenman G: The passive role of nasal septal cartilage in mid-facial growth. Plast Reconstr Surg 41:536, 1968

Moyers RE: Handbook of Orthodontics, 3rd ed. Chicago, Year Book, 1973

Sarnat BG, Muchnic H: Facial skeletal changes after mandibular condylectomy in growing and adult monkeys. Am J Orthod 60:33, 1971

Sarnat BG, Wexler MR: Growth of the face and jaws after resection of the septal cartilage in the rabbit. Am J Anat 118:755, 1966

Scott JH: Growth at facial sutures. Am J Orthod 42:381, 1956

Servoss JM: An in vivo and in vitro autoradiographic investigation of growth in synchondrosal cartilages. Am J Anat 136:479, 1973

Sicher H: Oral Anatomy, St. Louis, CV Mosby Co, 1952

Stockli PW, Wilbert HG: Tissue reactions in the temperomandibular joint resulting from anterior displacement of the mandible in the monkey. Am J Orthod 60:142, 1971

Wexler MR, Sarnat BG: Rabbit snout growth after dislocation of nasal septum. Arch Otolaryngol 81:68, 1965

Young RW: The influence of cranial contents on postnatal growth of the skull in the rat. Am J Anat 105:383, 1959

· 7 ·

Growth Prediction

Year after year behold the silent toil
That spread his lustrous coil;
Still, as the spiral grew,
He left the past year's dwelling for the new,
Stole with soft step its shining archway through,
Built up its idle door,
Stretched in his last-found home, and knew the old no more.
        Oliver Wendell Holmes, *The Chambered Nautilus*

MOTIVATION

Initially, in any field of study (economics, physics, political science, etc.) fragmented bits of information develop into an accepted body of knowledge. Soon, laws, theorems, and principles begin to appear, interspersed with theories and hypotheses. As soon as confidence is gained in the universality of some of the tenets of the subjects (sometimes unwarranted), bolder protagonists begin to extrapolate and even predict.

The presumption that one's information has sufficiently crystallized to embody predictive properties is a natural consequence of man's inquisitive nature. If he has discovered a "truth," then it must surely apply to all things at all times. A hypothesis tested a hundred times, a thousand times, and then a million times, always with the same results, becomes a law. At this point, we accept that the million and first time the law is tested will give the results as the preceding million. If we can find no exception to the truth, then we are justified in predicting results or events based on our newly discovered law.

Craniofacial biology has accumulated a sizable body of facts and theories and, inevitably, as with other disciplines, has recently begun to test the waters of prediction. Some scientists in the field are seeking truths about human facial growth that will allow us to extrapolate an adult face from the face of a child.

The cost of accumulating such expertise will be years of diligent research (if it is possible at all), and you might wonder what the payoff is for the effort expended. Primarily, craniofacial biologists just want to know, but being able to predict the growth of the head would be beneficial in at least two respects. In the first instance, being able to foretell growth would verify that we understand a great deal about it; the science of prediction would demand that our body of knowledge be valid and cohesive. Second, the clinician dealing with the interception and/or correction of dentofacial malocclusions would be helped immeasurably if he could predict with a degree of certainty the adult features of his patient. If the dentist could forecast the child's adult appearance with or without therapeutic intervention, decisions could be made about the timing, type, and length of treatment. In borderline cases, decisions on whether to treat at all could be better weighed.

Having accepted that such knowledge would be intellectually satisfying and clinically useful, craniofacial biologists have begun to probe from divergent points of view. One group has taken the basic approach, seeking universal laws of growth applicable to the human, while other groups have sought the techniques for prediction using case histories and clinical trial and error. Reassuringly, the two approaches are beginning to merge.

For growth prediction to be feasible at all, craniofacial growth must be orderly. If it were irregular, random, or discountinuous, then no amount of effort could develop a system or mathematical formula to predict it. Fortunately, there is every indication that it is orderly. In fact, Moss and Salentijin have proposed a mathematic expression to describe all human mandibular growth, and Ricketts has created a cephalometric technique for drawing an arc down which the mandible can be expected to grow.

But before we get into the specifics, let us explore the foundations that have supported these developments. The mathematical model that describes mandibular growth has evolved from D'Arcy Thompson's study of seashells and Moss's interrelated theories of the functional matrix, gnomonic facial growth, and facial neurotrophism.

GNOMONIC GROWTH AND THE LOGARITHMIC SPIRAL

The Chambered Nautilus

In his classic book, On Growth and Form, Thompson brilliantly analyzed the growth of certain sea shells, not only classifying them according to their patterns of enlargement, but in many instances developing equations to fit the process.

For our purposes, the discussion will be restricted to the chambered nautilus. This beautiful shell, pictured in Fig. 7.1, has two fundamental characteristics of importance to us. First, the shell grows in size but does not change its shape. Although the shell grows asymmetrically (adds new mineral salts at one end only), the original shape remains constant (Fig. 7.2). This feature would qualify the nautilus as one of those things described by Aristotle that suffers no

FIGURE 7.1   X-ray of a nautilus. Note the gnomonic pattern in the addition of chambers.

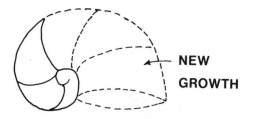

FIGURE 7.2   Diagram of new growth (gnomonic) of a nautilus, which enlarges the shell while maintaining the same configuration.

alteration except magnitude when they grow. That portion or increment which when added achieves such a growth is called in Greek a *gnomon*. The process of growth whereupon the addition of a figure or body leaves the resultant figure or body similar to the original is called gnomonic growth.

The second characteristic of the nautilus is that its gnomonic growth can be described by a particular kind of curve, the logarithmic or equiangular spiral (Fig. 7.3).

This spiral is characterized by the movement of a point away from the pole along the radius vector with a velocity increasing as its

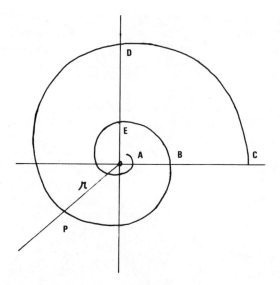

FIGURE 7.3   An equiangular or logarithmic spiral. $\angle A = \angle B = \angle C$, and $\angle E = \angle D$, r = radius vector.

distance from the pole. If you begin to swing a tethered ball around your head and gradually and steadily let the line slip out, the arch traversed by the weight will be an equiangular spiral. The "equal angles" of the spiral are *A*, *B*, and *C*, or *D* and *E*, or any formed by a straight line drawn out from the pole. Because the radius vector increases its length in geometrical progression as it sweeps through successive equal angles, the equation of the spiral will be $r = a^\theta$. (r = distance *OP*, θ = vectorial angle, a = a constant).

From algebra, when $m^x = n$, x is called the logarithm of n to the base m. Thus, the equation $r = a^\theta$ may be written $\log r = \theta a$, or since a is a constant, $\theta = k \log r$. This means that the vector angles about the pole are proportional to the logarithms of the successive radii. Consequently, the name *logarithmic spiral* has become popular.

Thus, to the spiral of the nautilus has been fitted a precise logarithmic formula. Since future growth of the animal will continue along that curve, the spiral can be generated at any time to reveal the final shape.

What a powerful tool it would be to have similar mathematic expressions fitted to human craniofacial growth! Nature has obviously cast a tight genetic mold for the nautilus, but is such orderliness restricted to shelled creatures? Probably not. A dramatic about face is not the way of nature. Did evolution cease to use gnomonic growth for higher forms millions of years ago in the vast reaches of the oceans? Recent evidence suggests that it did not—the structures that grow gnomonically in the human head are just less apparent.

Gnomonic Growth of Rectilinear Figures

The property of continued similarity exists in no plane curve other than the logarithmic spiral, but many rectilinear figures can be shown to possess this property. Consider the cone (Fig. 7.4). No matter how the original cone (shaded) is added to (one, two, or three sides) the resulting figure has the same shape. This is not unexpected, since the nautilus is but a cone rolled up and represents an especial case of cone B.

Other figures possess the ability to demonstrate a logarithmic spiral growth if the increments added to them are gnomons. Consider a rectangle A (Fig. 7.5) with the dimensions of the sides at a ratio of $1:\sqrt{2}$. If we now add an identical rectangle B, the combined shape (A + B) is similar to A, since $1:\sqrt{2}$ as $\sqrt{2}:2$ ($1/\sqrt{2} = \sqrt{2}/2 = 0.707$).

Rectangle B is a gnomon to rectangle A. If we continue to add rectangles of such dimensions that the new combined areas are consis-

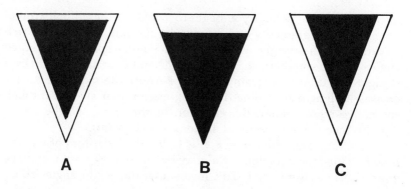

FIGURE 7.4 A triangle is a figure (rectilinear) that maintains its configuration despite the number of sides to which area is added. A has been added to on three sides, B one side, and C two sides. Adapted from D.W. Thompson: On Growth and Form, 2nd ed. Oxford, England, Oxford University Press, 1952.

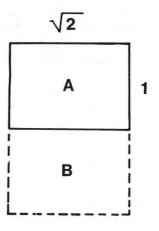

FIGURE 7.5 A rectangle with the dimensions of the sides at a ratio of $1:\sqrt{2}$ is a rectilinear figure.

tently proportional to the others (A: A + B : A + B + C : A + B + C + D) (Fig. 7.6), a configuration is obtained which can be circumscribed by an equiangular spiral.

Triangular shapes exhibit similar properties, as can be demonstrated by adding a series of gnomons to an isosceles triangle, converting it into larger and larger triangles, all similar to the first. The apices of these triangles have their loci on an equiangular spiral (Fig. 7.7).

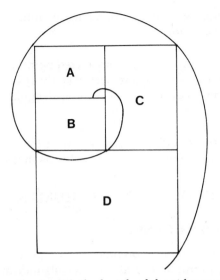

FIGURE 7.6 A rectangle with the length of the sides at a ratio of $1:\sqrt{2}$ exhibits gnomonic growth when (1) area is added that equals the dimensions of the already existing areas and (2) the area is added in such a way as to preserve the ratio of the sides at $1:\sqrt{2}$. Adapted from D.W. Thompson: On Growth and Form, 2nd ed. Oxford, England, Oxford University Press, 1952.

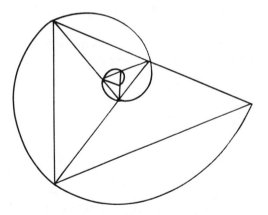

FIGURE 7.7 An array of isosceles triangles is created by adding a series of gnomons to the appropriate sides. The apexes of these triangles have their loci on a logarithmic spiral. Adapted from D.W. Thompson: On Growth and Form, 2nd ed. Oxford, England, Oxford University Press, 1952.

To Thompson we owe the following definition, which sums up the preceding discussion:

> Any plane curve proceeding from a fixed point (or pole), and such that the vectorial area of any sector is always a gnomon to the whole preceding figure, is called an equiangular, or logarithmic spiral.

If such relationships could be discovered in the face, then predictions about its growth would be as feasible as in the nautilus.

GNOMONIC GROWTH OF THE HUMAN HEAD

Growth of Craniofacial Spaces

What do we know about gnomonic growth in the face? Moss has pioneered study in this area, and early results indicate that orofacial capsular matrices, particularly the oronasopharyngeal functioning spaces, manifest gnomonic growth. In one study, the heads of human fetuses with crown–rump lengths ranging from 26 to 163 mm were sectioned midsagittally for direct measurements of the oral, nasal, and pharyngeal cavities (Fig. 7.8). From this analysis, the gnomonic growth of the nasal (*left*) and oral (*right*) functioning spaces is depicted as its occurs in fetal man. The numbers represent the crown–rump length. Notice how the spaces enlarge but do not change shape.

Moss has cited the earlier work of Burdi as the finest example of gnomonic growth of the nasal functioning space. Using the cephalo-

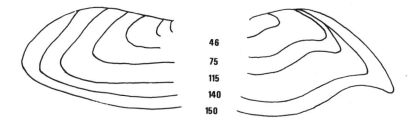

FIGURE 7.8   The nasal functioning spaces of human fetuses of various crown–rump lengths (*left*). The oral functioning spaces of the same fetuses (*right*). Both areas manifest gnomonic growth. Adapted from L. Salentijn and M.L. Moss: Morphological attributes of the logarithmic growth of the human face: gnomonic growth. Acta Anat 78:185, 1971.

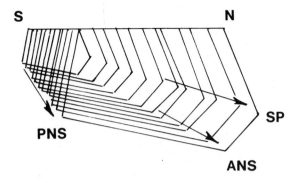

S   N

SP

PNS

ANS

FIGURE. 7.9 Gnomonic growth of the human fetal nasal space during the
second trimester. S = sella, PNS = posterior nasal spine, N = nasion,
ANS = anterior nasal spine, SP = septal point. Adapted from A.R.
Burdi: Sagittal growth of the naso-maxillary complex during the second trimester of the human development. J Dent Res 44:112, 1965.

metric points sella (S), nasion (N), anterior and posterior nasal spines
(ANS, PNS), and septal point (SP), Burdi developed the representation of the growth of the human fetal nasal space during the second
trimester shown in Fig. 7.9. Notice how this space has resituated the
surrounding skeletal units in such a way that the overall shapes remain astoundingly similar. Such orderliness should possess some
predictive value.

From Burdi's data, Moss constructed a new figure for
crown–rump lengths of 100, 300, and 400 mm (Fig. 7.10). We have
taken the liberty of dashing in the presumed dimensions of a
crown–rump length of 200 mm. Additionally, we have drawn over
the figure a possible logarithmic spiral suggested by the gnomonic
growth.

Ricketts has suggested a number of gnomonic figures that are
correlated to three branches of the trigeminal nerve—the opthalmic
(V_1), maxillary (V_2), and the inferior alveolar (V_3)—a branch of the
mandibular nerve. V_1 passes through the superior orbital fissure to
supply sensory innervation to the eye and upper face. V_2 exits the
cranium through the foramen rotundum, supplying sensory nerves
to the maxillary complex. V_3 enters the ramus at the mandibular
foramen, providing sensory innervation for the lower dentition (Fig.
7.11A) Ricketts considers that these nerves are quite important to the
organization of facial morphology, a tacit recognition of the neurotrophic concept, which will be discussed in the next section. The
gnomonic figures he visualizes are shown in Fig. 7.11B.

The focal point for the region supplied by one of these branches

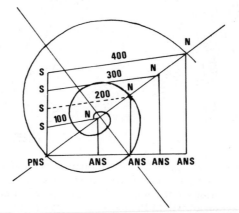

FIGURE 7.10   The logarithmic spiral suggested by the gnomonic growth of the human fetal nasal space during the second trimester. Adapted from L. Salentijn and M.L. Moss: Morphological attributes of the logarithmic growth of the human face: gnomonic growth. Acta Anat 78:185, 1971.

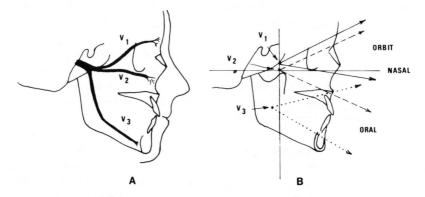

FIGURE 7.11   (A) V_1, V_2, and V_3 are branches of the trigeminal nerve as they would be viewed from their foramina in the x-ray. (B) A composite of 5-year-old children showing the facial gnomonic figures. The orbit angle has its vertex at V_1, the lower border of the superior orbital fissure. V_2, the vertex of the nasal angle, orients on the foramen rotundum, and V_3, the vertex of the oral region, on the mandibular foramen. Adapted from R.M. Ricketts: The value of cephalometrics and computerized technology. Angle Orthod 42:179, 1972.

is the foramen associated with the nerve. For example, when lines are extended from the superior orbital fissure base, the content of the orbit appears to expand within a constant angle (Fig. 7.12 *top*). This property, i.e., increase in size without change in shape, satisfies the definition of gnomonic growth. The nasal angle A–Pt–ANS (where V_2 is the vertex) describes growth of the nasal capsule (Fig. 7.12 *middle*). The oral cavity enlarges within the angle ANS–Xi–Po created by locating the vertex at Xi, a point that represents the mandibular foramen or V_3 (Fig. 7.12 *bottom*).

Moss and Salentijn, essentially in agreement with Ricketts, have concluded that the orofacial capsule responsible for the translative movement of the mandible creates a gnomonic growth. They have reached this determination following their analysis of the foramina of the skull and mandible through which the inferior alveolar nerve passes. Why they chose to examine pathways for innervation as examples of logarithmic spirals requires a momentary digression.

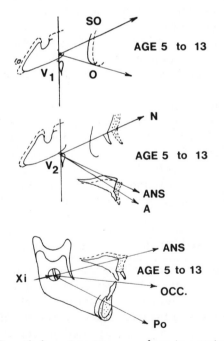

FIGURE 7.12   Growth from 5 to 13 years of age in a series of 40 children. (*Top*) Growth from the orbital vertex (V_1); (*Middle*) expansion of the nasal cavity from V_2; (*bottom*) the gnomonic growth of the oral cavity from the perspective of V_3. Adapted from R.M. Ricketts: The value of cephalometrics and computerized technology. Angle Orthod 42:179, 1972.

Neurotrophism

The functional matrix theory disclaims any intrinsic genetic determination by bone cells, and therefore the information about the rate and limitation of growth must exist somewhere in the capsules. Moss contends that to a great extent the messages necessary for controlling growth are derived from the nerves that innervate these capsules. Ultimately, the DNA that dominates craniofacial growth resides in the chromosomes of brain cells. RNA or other messages are carried to end organs by axoplasmic flow, a process first shown by Weiss.

The classic experiments that demonstrated the role of nerves in controlling growth involved the regeneration of limbs of salamanders. When a leg of this vertebrate is amputated and the nerve supply left intact, a new limb will develop in a matter of weeks. If, on the other hand, the nerve supply should happen to be interrupted at the time of amputation, only primitive healing will ensue. Many other experiments could be cited that demonstrate the trophic ("to nourish") influence of the nerve. This function of the nerve should be distinguished from conduction, an electrochemical process.

Assigning such a prominent role to the process of neurotrophism, or neural nourishment, naturally dictates that at no time are the nerves to be subjected to torsion, compression, tension, or shear. With respect to the inferior alveolar nerve, it is absolutely imperative that, from its exit out of the foramen ovale as the fifth cranial nerve, to its entry into the mandibular foramen, along the mandibular canal and out the mental foramen, its integrity never be threatened by growth or functioning of the lower jaw. As a vital constituent of the orofacial capsule, Moss found it reasonable to speculate that the pathway of the inferior alveolar nerve follows a logarithmic spiral.

Logarithmic Growth of the Human Mandible

Craniometric studies were performed on American Indian skulls of various ages. With small lead shots affixed to the foramen ovale, mandibular foramen, and mental foramen, lateral x-rays effectively outlined the pathway of the inferior alveolar nerve. In Fig. 7.13A are drawn four mandibles with the three neural foramina accented by the radiopaque shot. They are representative of mandibles with fetal, deciduous, mixed, and adult dentitions. On the right, these foramina are aligned on a curve that fits them all (Fig. 7.13B). This curve is a logarithmic spiral for which the equation is $\log r = \log a + ko$, where r = length of a radius, o = angulation of this radial line measured in

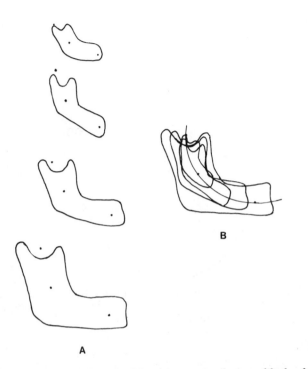

FIGURE 7.13   (A) The location of the foramen ovale, mandibular foramen, and mental foramen in the skulls of four ages—fetal, deciduous, mixed, and adult. (B) The foramina are perfectly aligned on a single curve, a logarithmic spiral. Adapted from M.L. Moss and L. Salentijn: The logarithmic growth of the human mandible. Acta Anat 77:341, 1970.

radians, and a = value of r when $o = 0$. The curve in Fig. 7.14 has a value of 1.91 for log a and $k = -0.189$. Consequently, it is possible to generate the curve representing human mandibular growth at any time.

At this juncture, it is important to point out that the mandible does not actually grow in space as pictured in Fig. 7.13B. During all stages of development, the corpus stays in essentially a horizontal position. At the same time, the mandible moves down the logarithmic spiral course of the inferior alveolar nerve. For these two statements to be compatible requires that the entire spiral be able to rotate. Looking from the right side, the spiral would be pivoted backward early on (Fig. 7.15). As the mandible increased in size, the whole spiral would rotate clockwise and the corpus would remain horizontal.

5 cm

├─────┤

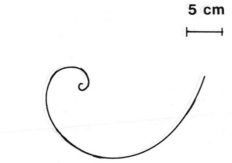

FIGURE 7.14 The logarithmic spiral formulated by Moss, which coincides with the three foramina of the inferior alveolar nerve and which describes the path of mandibular growth.

MANDIBLE CHANGES
POSITION ON SPIRAL
AS IT INCREASES
IN SIZE

ROTATION OF SPIRAL
WITH AGE

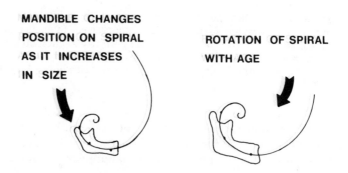

FIGURE 7.15 As the foramina separate during growth, the mandible continually assumes a position where there is less curvature of the spiral (*left*). Because the mandible does not actually grow up and out, rotation of the spiral must occur (*right*).

ARCIAL GROWTH OF THE MANDIBLE

Definition and Landmarks

Ricketts, using a trial and error procedure with longitudinal cephalometric records and computers, has developed a method to determine the arc of growth of the mandible. Not only does this arcial analysis lend itself to prediction, but many important biologic and clinical implications have arisen secondarily. In essence, the principle is:

> A normal human mandible grows by superior–anterior (vertical) apposition at the ramus on a curve or arch which is a segment formed from a circle. The radius of this circle is determined by using the distance from mental protuberance (Pm) to a point at the forking of the stress lines at the terminus of the oblique ridge on the medial side of the ramus (point Eva).

To understand the rationale behind the selection of these points and the technique for contructing the arc requires a stepwise development of Ricketts' thinking. The effort required to follow this process is well justified, because his view of mandibular growth adds new dimensions to dental craniofacial biology.

Improved visualization of the condyle and coronoid process by body-section roentgenography (laminagraphy) enabled Ricketts to better observe the "bending" of the mandible from infancy to maturity. Bjork, in a previous study that included metallic intraboney implants for cephalometric analysis, showed extensive variation in mandibular bending and also revealed lower angular border resorption to be a typical phenomenon in the normal patient.

The insight into mandibular growth augmented by the surface analysis for patterns of apposition and resorption by Enlow eliminated the mandibular plane as an acceptable reference base. The next move was to identify a "central core" cephalometrically that persists immune to the vagaries of drift. This internal structure was thought to have some relationship to the inferior alveolar nerve, and for this reason the mandibular foramen was selected. Unfortunately, lateral roentgenographic cephalometry does not reveal the landmark with certainty. To overcome this drawback, a new point Xi and a technique for determining its position were developed (Fig. 7.16). Point Xi represents the centroid of the ramus and is the crossing point of the diagonals drawn from the corners of a rectangle (for details, see Ricketts*). This point very successfully locates the opening

---

* Ricketts RM: A Principle of Arcial Growth of the Mandible. Angle Orthod 42:368, 1972.

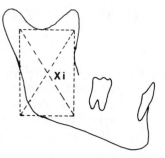

FIGURE 7.16   Xi represents the geometric center of the ramus, determined by the intersection of the diagonals of the rectangle drawn to fit the width and height of the ramus at its narrowest dimensions.

of the mandibular canal. (Note the convergence with the neurotrophic concepts of Moss.)

The second point Pm (protuberance menti) was selected because it is an identifiable and stable landmark (Fig. 7.17). Dc, the third point, represents the bisection of the condyle neck as high as visible in the cephalometric film below the fossa. A line from Dc to Xi constitutes the condyle axis and the one from Xi to Pm the corpus axis. Consequently, these planes can be studied for dimensional and angular changes.

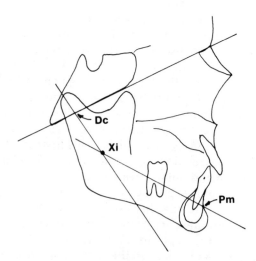

FIGURE 7.17   The lines Dc–Xi (condyle axis) and Xi–Pm (corpus axis) describe the "core" of the mandible.

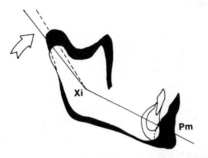

FIGURE 7.18 Mandibles on an "average" bend one-half degree per year using the Dc–Xi–Pm axis. Adapted from R.M. Ricketts: A principle of arcial growth of the mandible. Angle Orthod 42:368, 1972.

The Arc of Growth

When an "average" mandible at time 1 was superimposed on an "average" mandible at time 2, using Xi and the corpus axis as reference, the mandibles were found to bend about one-half degree each year (Fig. 7.18). (Average means a composite made from a computer printout).

The next step involved a trial and error format that finally led to the development of the arc that described this bending. (*Note:* This arc is not universal in describing all human mandibular growth as does the logarithmic spiral of Moss. In Ricketts' analysis, each individual generates his own unique arc).

Three curves were ultimately drawn (A, B, and C of Fig. 7.19) before C was determined to best fit the true arc of growth of the mandible.

Curve A, which passed through Dc, Xi, and Pm, did not produce enough bending, so that in the prediction of growth over a sufficient time span, the resulting mandible would be too obtuse.

Curve B, passing from the tip of the coronoid process, touching the anterior border of the ramus at its deepest curve and through the same Pm point, constituted a segment of a circle with too small a radius. The projected mandible was bent excessively.

These two unsuccessful arcs obviously bracketed the true arc, which must be somewhere in the mandible between the condyloid and coronoid processes and between Xi and the anterior border of the ramus. Analysis of the stress lines of a very old skull revealed an area at almost the center of the upward and forward quadrant of the ramus on the lingual surface that appeared at the confluence of various fields of force (Fig. 7.20).

The lateral surfaces also exhibited a Y-shaped convergence of

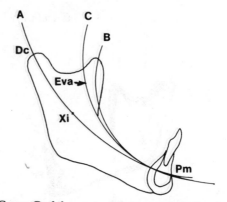

FIGURE 7.19   Curve C of three possible curves best fits the arcial growth of the mandible. Adapted from R.M. Ricketts: A principle of arcial growth of the mandible. Angle Orthod 42:368, 1972.

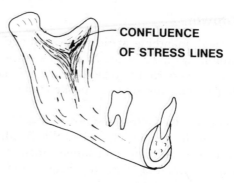

FIGURE 7.20   Point Eva, the confluence of stress lines on the medial surface of the ramus.

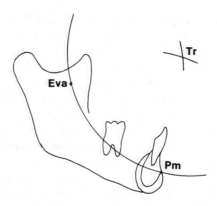

FIGURE 7.21   Tr represents the intersection of two arcs with their radii equal to Eva–Pm. Eva constitutes the center of one arc, Pm the other. The curve created through Eva and Pm with Tr as center represents the arc of mandibular growth.

several stress lines, and Ricketts reasoned that these areas were important in mandibular growth. The medial confluence was given the name Eva, in honor of his mother.

When a new point Tr is used as the center of a circle whose segment passes through Eva and Pm and that has a radius equal to Eva–Pm, the true arc for growth of the mandible is developed (Fig. 7.21). Computer analysis revealed that the predicted mandible was almost absolutely correct in size and form when compared with the final composite.

Having satisfied himself with the accuracy of the arc, Ricketts determined the yearly increments necessary to complete the forecast of growth. Annual increases of 2.5 mm were, when averaged over the years of time, an excellent population constant. Growth was found to cease at 14.5 years for females and 19 for males.

This technique now represents a convenient and fairly reliable method for mandibular growth prediction, although several disclaimers have questioned its accuracy. Growth in other areas of the skull can also be forecast, but the methodology does not involve the generation of curves. Consequently, there is no convergence with the philosophy of Moss.

THE MANDIBLE—CONTINUED

Repercussions of Arcial Growth

We now have adequate background to analyze the ramifications of the view that mandibular growth is somehow associated with a curve. We will focus our attention on the arcial growth pattern established by Ricketts, since he has studied the consequences most fully, but bear in mind that other curves, including the logarithmic spiral, should generate similar findings.

Figure 7.22 depicts Rickett's concept of arcial mandibular growth, and when compared to the conventional orientation along the mandibular plane (Fig. 4.26), important differences are immediately apparent. Fundamentally, arcial growth is dependent on superior–anterior growth of the ramus, rather than posterior growth, which would be required of the type of growth pictured in Fig. 4.26. Clearly, the increase in size involved a vertical and not a horizontal process.

This shift in the direction of growth automatically alters the mechanism whereby room is made for the second and third molars. In the conventional view, major resorption is required of the anterior border of the ramus to provide eruption space for the distal teeth.

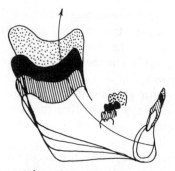

FIGURE 7.22 When serial tracings are aligned on the arc of growth and
registered at the pogonion and the anterior border of the ramus at
the coronoid crest, the vertical apposition at the superior border of
the ramus is displayed. Adapted from R.M. Ricketts: A principle of
arcial growth of the mandible. Angle Orthod 42:368, 1972.

The arcial growth pattern obviates the need for such resorption, because the lower first molar tends to erupt upward and forward with
the occlusal plane. Thus, the "rolling" pattern of mandibular growth
deemphasizes the need for bone removal. As a corollary to the
movement of the lower arch, it becomes easier to visualize how ankylosis of a tooth can lead to the degree of submergence that is occasionally seen.

The arcial movement of the mandible also pushes the symphysis, or chin, under the denture as the teeth erupt upward and
forward. This explains the development of the "chin button" despite
minimum apposition in this area.

These are but a few of the important offshoots inherent to an
acceptance of the arcial principle. As a intellectual pursuit only, it
does not matter to the mandibles of the world how we view their
growth. But beyond theory, important clinical decisions are currently being made predicated on the reality of arcial growth. For instance, anchorage of the lower arch during orthodontic treatment is
improved by positioning the roots of the molars buccal under cortical bone, which prevents upward and therefore forward eruption of
the whole lower dental arch.

Arcial Growth in Space

Before ending the discussion on the arcial growth of the mandible,
attention should be drawn to the actual position of the arc in space.
Consider the following tracing in Fig. 7.23 and particularly note the
facial axis, which connects Pt with the cephalometric gnathion (Gn).

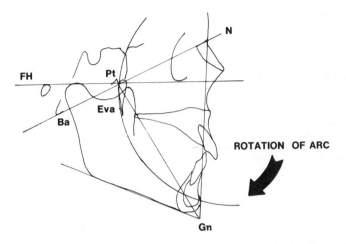

FIGURE 7.23   As the mandible grows down its arc anteriorly, the entire arc must rotate posteriorly to allow for vertical growth of the lower face.

The angle Ba–Pt–Gn is nearly always 90° and is virtually stable during the entire growth process. This finding is not consistent with what would occur if the upward and forward growth of the ramus rigidly followed the arc and were allowed to shift the chin upward and forward without compensation somewhere.

As the arc develops, there must be a rotation (clockwise from the right lateral view) in order to maintain the facial axis at a near constant. This rotation is a close parallel to the rotations of the logarithmic spiral, and the pivot points are presumably related to the neurotrophic bundle that supplies the mandible.

BIBLIOGRAPHY

Bjork A: Prediction of mandibular growth rotation. Am J Orthod 55:585, 1969

Burdi AR: Sagittal growth of the naso-maxillary complex during the second trimester of human prenatal development. J Dent Res 44:112, 1965

Enlow DH, Harris DB: A study of the postnatal growth of the human mandible. Am J Orthod 50:25, 1964

Greenberg LZ, Johnston LE: Computerized prediction: The accuracy of a contemporary long-range forecast. Am J Orthod 67:243, 1975

Hirschfield WJ, Moyers RE: Prediction of craniofacial growth: The state of the art. Am J Orthod 60: 435, 1971

Moss ML: Neurotropic processes in orofacial growth. J Dent Res 50:1492, 1971

Moss ML: Neurotrophic regulation of craniofacial growth. In McNamara JA Jr (ed): Determinants of Craniofacial Form and Growth. Ann Arbor, University of Michigan, 1975.

Moss, ML, Salentijn L: The logarithmic growth of the human mandible. Act Anat 77:341, 1970

Moss ML, Salentijn L: The unitary logarithmic curve descriptive of human mandibular growth. Acta Anat 78:532, 1971

Ricketts RM: The evolution of diagnosis to computerized cephalometrics. Am J Orthod 55:795, 1969

Ricketts RM: The value of cephalometrics and computerized technology. Angle Orthod 42:179, 1972

Ricketts RM: A principle of arcial growth of the mandible. Angle Orthod 42:368, 1972

Ricketts RM, Bench RW, Hilgers JJ, Shulhof R: An overview of computerized cephalometrics. Am J Orthod 61:1, 1972

Salentijn L, Moss ML: Morphological attributes of the logarithmic growth of the human face: gnomic growth. Acta Anat 78:185, 1971

Thompson DW: On Growth and Form, 2nd ed. Oxford, England, Oxford University Press, 1952

Weiss P, Hiscoe HB: Experiments on the mechanism of Nerve Growth, Exp Zool 107:315, 1948

· 8 ·

Development of
the Dentition

The fathers have eaten a sour grape, and the children's teeth are set on edge.
Jer. 31:29-30

The development of the dentition is an integral part of craniofacial growth, and the relationship of the teeth to each other, to their supporting structures, and to the face, both in static and dynamic situations, have motivated a great deal of research.

The formation, eruption, exfoliation, and exchange of teeth, from crowded jaws into an oral cavity subject to multiple forces, from positions deep within bone into an environment of changing jaw relationships, is one of the most fascinating subtopics of craniofacial biology.

TOOTH ERUPTION

Root Formation

While we can describe where the teeth erupt and all the associated histologic activity, we still do not understand the mechanism of tooth eruption. We do not know where the forces are generated that push or draw the tooth through millimeters of bone and, in the case of succedaneous teeth, through the roots of their predecessors. There are numerous theories that explain the eruption mechanism, and nearly all of them revolve around some aspect of root formation.

While it is not in the purview of this text to elaborate on dental embryology, a brief account of root and periodontal ligament formation might clarify the discussion of tooth eruption.

Both the primary and permanent teeth have a similar developmental history, including the following stages: (1) lamina; (2) bud; (3) cap; (4) bell; and (5) appositional. While in the bell stage, the enamel organ differentiates into four cell types: (1) inner enamel epithelium, adjacent to the dental papilla; (2) stratum intermedium, the layer of cells overriding the inner enamel epithelium; (3) stellate cells, comprising the bulk of the enamel organ; and (4) outer enamel epithelium. The cells of the inner enamel epithelium become ameloblasts following differentiation of dental papilla cells into odontoblasts. The appositional stage begins with the deposition of enamel (E) and dentin (D) as the ameloblast and odontoblasts withdraw from each other and the dentinoenamel junction (Fig. 8.1).

The outer enamel epithelium becomes discontinuous, allowing entry of cells from the surrounding dental sac, while the stellate reticulum is withdrawn to make room for the crown. The rim of the

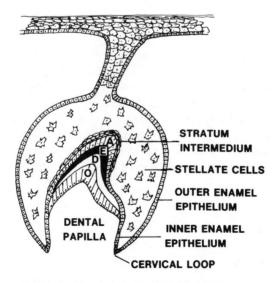

FIGURE 8.1   The beginning of enamel and dentin formation in the appositional stage; (A) ameloblasts, (O) odontoblasts, (E) enamel, and (D) dentin.

enamel organ, the cervical loop, is composed of two layers—the inner and outer enamel epithelium.

When enamel formation is complete, the crown is fully formed and root development is begun. Just before the ameloblasts at the junction of the inner and outer enamel epithelium deposit their matrix, a proliferation of cells in this area lengthens the cervical loop. This structure, now called Hertwig's epithelial root sheath, determines the number, size, and shape of the roots.

Dentin formation continues uninterrupted from the point where the enamel terminates, but the matrix is deposited against the root sheath instead of ameloblasts (Fig. 8.2). On the roots, the dentin is covered by cementum rather than enamel. The deposition of cementum follows the invasion of cemontoblasts, a phenomenon that is allowed when the continuity of the sheath is interrupted by dentin calcification.

The cementoblasts are differentiated mesenchymal cells arising from the primitive cells of the dental sac, or follicle, which surrounds the developing tooth. This tissue is designed to eventually form the periodontal ligament, which by its attachment of fibers to alveolar bone and cementum, anchors the teeth in the jaws. Part of the collagen fibrils secreted by the cells of the primitive periodontal ligament becomes incorporated as Sharpey's fibers into the alveolar bone; other fibrils become embedded into the cementum (cemental fibers). According to some authorities, these two attached portions are then linked by a middle zone of fibers called the intermediate plexus (Fig. 8.3)

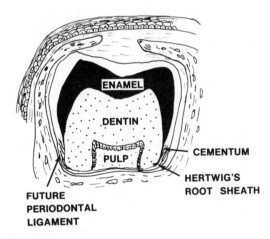

FIGURE 8.2   The initiation of root formation, deposition of cementum, and elaboration of the periodontal ligament.

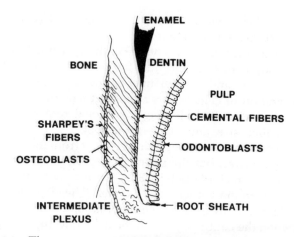

FIGURE 8.3   The components of the developing periodontal ligament—
Sharpey's fibers, cemental fibers, and the intermediate plexus.

The periodontal ligament develops successively from the dentinoenamel junction toward the apex, deriving its components from the active growth area lateral to the root end. A random formation of collagen fibers, initially seen in this area, is soon followed by an orientation of fibers coronally, with their origin in the cementum. Development of the intermediate plexus completes a vertically inclined ligament.

Theories of Eruption

Apparently, the tooth bud is incapable of any movement until the root begins to form. In someway, then, the initial eruption of teeth is associated with root development, but be careful to note the word *initial.* Experience has shown that teeth can erupt fully with incomplete root formation, that some teeth normally erupt through a distance that exceeds their root length, and some impacted teeth with totally formed root are capable of eruption once the obstructions are removed.

In Fig. 8.4 are diagrammed a number of the possibilities to explain the correlation between root growth and tooth eruption:

1. Sicher proposed one of the earliest theories of tooth eruption after he observed at the apical end of the growing root a fibrous thickening composed of an intermingling of pulp, dental sac, and new periodontal fibers. He visualized a continuous, fibrous, ham-

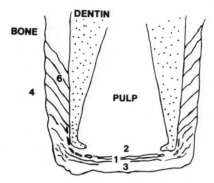

FIGURE 8.4 Sites of actions for the leading theories of tooth eruption: (1) cushioned hammock ligament, (2) site of fluid accumulation, (3) fundic bone growth, (4) bone wedging action, (5) tissue fluid gradient, (6) periodontal ligament.

mocklike ligament, which when imbued with extracellular fluid becomes a "cushioned hammock ligament." This structure supposedly acts as the barrier to apical pulpal growth, with the result that any expansion of root and pulpal tissue displaces the crown coronally. Most investigators disclaim the existence of this ligament.

2. Several investigators have proposed that higher tissue pressure or an accumulation of tissue fluid at the developing apex might force the hard tissues of the tooth and bone apart. However, the eruption of teeth with fully formed roots with minute apical openings weighs against this theory.

3. Fundic bone growth could propel the tooth coronally, but earlier in this text it was noted that bone does not grow well into the face of direct pressure. This theory proposes a property of bone that it does not have.

4. Another postulated bone mechanism ascribes eruption to the selective remodeling of bone whereby the tooth is "wedged" toward the oral cavity. This theory not only credits bone with the capacity to exert pressure but also with sufficient genetic information to govern such a process. Additionally, it is hard to envision a wedging action on a multirooted tooth with widely divergent roots.

5. By lowering the tissue fluid pressure above the tooth, a gradient of pressure could be created between the crown and the apex. This difference might effect a coronal movement of the tooth. Such a gradient has been shown to exist in animal studies, but experiments to modify the eruption rate by altering the fluid pressure have not been convincing.

6. The theory currently in vogue centers around the activity of the periodontal ligament. As the collagen fibers of the intermediate plexus mature, they are hypothesized to contract. Because the ligament is vertically inclined, such shortening would draw the tooth coronally. Continual turnover in the plexus would remove mature, contracted fibers and replace them with young elongated fibers that make new attachments to Sharpey's and cemental fibers. This theory is attractive because it requires some root formation and consequently some periodontal ligament, to be present before a tooth bud initiates movement. But additionally, it explains why a partly formed tooth or a completely developed buried tooth can still erupt.

The Gubernaculum Dentis

During the early stages, a primary tooth shares a bony crypt with its successor, but movements concurrent with the root development of the primary tooth and the apposition of alveolar bone results in the permanent tooth being situated inferiorly. The crypt of the permanent tooth becomes separated from the primary tooth by a bony ceiling. The isolation of the second tooth would be complete if it were not for a small communication between the gingival mucosa and the tooth follicle (Fig. 8.5). The pore (situated lingual to the primary teeth) contains connective tissue, blood vessels, nerves, and the

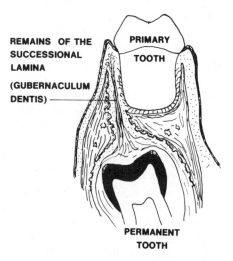

FIGURE 8.5 Relationship between the primary tooth, the gubernaculum dentis, and the permanent successor.

remains of the successional lamina. This aggregation of tissues (given the name *gubernaculum dentis* by Scott) has been implicated in two aspects of tooth eruption. First, it has been suggested that it might act to maintain the relative position of the follicle within the alveolar process. As the vertical dimension of the child's jaws increases with age, the alveolar processes deposit bone occlusally. Such activity would effect an ever submerging position of the teeth if it were not for the continual traction from the gubernaculum. Second, the structure might act as a guide or a path of least resistance for tooth eruption. Although a direct correlation between the location of the gubernaculum and the eruption path of the teeth has not been experimentally or clinically demonstrated, the consistency of the emerging position assumed by the permanent teeth strongly suggests a guidance mechanism.

Exfoliation of Primary Teeth

The replacement of the primary dentition is a multifactorial process. It is by no means a simple system in which eruption pressure elicits resorption of the primary tooth. Instead, phenomena such as a predilection to resorption by the roots of primary teeth and selective alveolar bone remodeling are intrinsic to the mechanism.

When the permanent tooth begins to erupt (for whatever reason), it is apparently guided by the gubernaculum. The pathway for the erupting anterior teeth is essentially lingual to their predecessors. This results in an oblique lingual pattern of resorption, manifested principally by the incisors.

The ceiling of alveolar bone is first removed followed by a sequential attack on the cementum and the dentin. Involved are bone osteoclasts and very similar cells that resorb the roots called odontoclasts. The resorption is not a continuous process, but is rather one of alternating periods of destruction and apposition. Even those primary teeth that have no succedaneous teeth under them exhibit an ebb and flow of resorptive activity. The fact that such teeth are almost invariably exfoliated, albeit somewhat later than normal, strongly suggests that roots of primary teeth have an intrinsic self-destruct mechanism. Pressure from the erupting permanent tooth only intensifies the inherent activity.

Removal of bone under the primary tooth is not the only site of an alveolar response. The supporting bone around the roots begins to resorb, and the cervical attachment drifts apically. The primary tooth is then losing attachment from two directions simultaneously—at the apical end by root resorption and from the coronal

end by bone loss. These converging activities expedite the exfoliation of the primary tooth.

This coronal loss of alveolar bone also lowers the bony support
of the emerging tooth to a functional level; the crown does not erupt
into a deep infrabony pocket. Consequently, the permanent tooth is
responsible for bringing up new alveolar bone with it as it erupts, a
process that allows new periodontal attachments to be made.

The alternating resorption and repair occurring at the roots or
primary teeth can sometimes encroach upon the periodontal ligament space (also in a state of change) and meld with the depository
activities of the bone. When the cementum and bone make direct
contact, the tooth is said to be ankylosed. This fusion may be so
miniscule as to be radiographically invisible but still sufficient to
prevent the primary tooth from drifting up with the alveolar bone
during vertical growth. The result is a submerging tooth and an in-
cipient malocclusion.

DEVELOPMENT OF THE HUMAN DENTITION

The development of the human dentition can be arbitrarily divided
into six stages:

1. Birth to the complete primary dentition
2. First intertransitional period
3. First transitional period
4. Second intertransitional period
5. Second transitional period
6. Adult dentition

These stages represent easily identifiable periods in what is actually
a continuum of activity. The development of crowns and roots is
largely hidden within the jaws, and only during the eruption or exchange of teeth is this activity manifested. There are, therefore,
seemingly quiescent interludes called intertransitional periods in
which the external appearance of the dentition remains unchanged.

Birth to the Complete Primary Dentition

At birth, the tooth buds of all the primary teeth are present and in
various stages of development. The incisors are somewhat crowded
at this time for two reasons: (1) the arches have not yet completely
rounded out anteriorly (only during the first 8 to 12 months are the

jaws capable of significant anterior apposition; thereafter, posterior growth and anterior displacement accounts for the increase in jaw size) and (2) the mechanism of formation of the incisors results in the very early attainment of the full mesiodistal width. (The development of the posterior teeth are more from the "inside out," and as a consequence, the follicles are slower in achieving their completed size.)

At 7 or 8 months postnatally, prior to the eruption of the lower primary incisor, all the teeth save the permanent second and third molars are present in some stage of development.

The succedaneous teeth assume specific locations—the replacements for the incisors and canines are apical and lingual, while the successors to the molars are initiated lingually but move labially to positions under the root bifurcations.

Usually, by the first year of age, sufficient jaw growth has occurred that the primary teeth are seldom crowded or overlapped. In fact, a normal and desirable dentition at this age will exhibit spacing.

Figure 8.6 illustrates the immaturity of the mandible of a child less than 1 year of age. A plane drawn along the occlusal surfaces passes through the condyle, clearly demonstrating the lack of ramal growth at this time. Either by genetic direction or lack of functional stimuli, the condyles are rudimentary. Remember that until the eruption of teeth, the tongue has been spread between the jaws in the typical suckling posture.

One cardinal feature of the primary dentition is the similarity of its orientation within the jaws compared to its position in the oral cavity. In contrast to the permanent teeth, they drop almost verti-

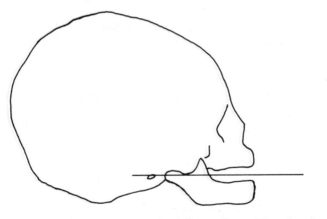

FIGURE 8.6 Lateral view of the skull of the newborn. Note that the occlusal plane passes through the immature condyle.

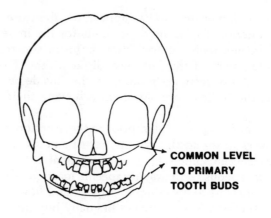

COMMON LEVEL TO PRIMARY TOOTH BUDS

FIGURE 8.7 The location of primary tooth buds in the jaws of an infant.

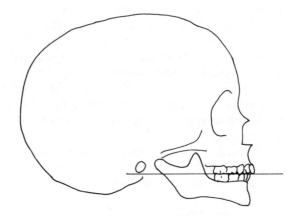

FIGURE 8.8 Lateral view of the skull of a two-and-a-half-year-old child. The occlusal plane is now located below the condyle.

cally into the mouth, requiring very little mesiodistal or buccolingual adjustment in their eruptive movement. In addition, the developing primary buds lie almost on the same occlusally oriented plane, a condition that might reflect the uniformity of the length of roots (Fig. 8.7).

The first primary tooth to erupt is usually one of the lower central incisors (between 6 and 8 months of age) followed by the upper centrals, upper laterals, and lower incisors. The first primary molars emerge around the fourteenth month, and their position, it has been suggested, determines the entire arch length–tooth size relationship.

This position wields such possible influence because the second primary molars and then the permanent molars in turn are limited in their mesial drift by its presence. It ultimately, then, might dictate the circumferential arch length from one first permanent molar to the other.

By the age 2½ years, the cuspids and second primary molars have erupted into occlusion. The entire face has undergone vertical growth as reflected by the superior positioning of the condyles above the occlusal plane (Fig. 8.8). The primary teeth are very upright, except for some labial inclination of the upper incisors. (The permanent incisors, which follow, have a pronounced labial inclination and the posterior permanent teeth assume a mesial tilt.) At this age, the roots of the primary cuspids and molars are not yet complete, and the crowns of the permanent centrals, laterals, cuspids and first molars are in various stages of formation.

First Intertransitional Period

The period between the completion of the primary dentition and the emergence of the first permanent teeth is marked by little obvious intraoral changes and multiple intrabone activities. According to Baume, the dentition of this period manifests several important characteristics. First, there is either spacing between the teeth from the time of eruption or there is not; teeth that enter the oral cavity without spacing do not separate with time. The spaced form he classifies as type I and the closed as type II. Often seen is a classic type of type I diastemata, called the primate spaces, mesial to the canines in the maxilla and distal to the canines in the lower arch.

Second, he pointed out that after the primary teeth are fully erupted, their sagittal and transversal dimensions are not physiologically altered. Only vertical changes ascribable to growth and occlusal attrition can be seen.

Third, the distal surface of the maxillary and mandibular second primary molars normally maintain either one or two vertical relationships during this period. If the surfaces are situated on the same vertical plane, they are described as having a flush terminal plane; if the distal surface of the lower molar is mesial to that of the upper, they are said to exhibit a mesial step. This relationship, while functionally unimportant at this time, can greatly influence the position of the first permanent molars later.

The arches by this time are incapable of any significant anterior growth, which means that space for the first permanent molars is achieved posteriorly by tuberosity apposition in the maxilla and ramal resorption in the mandible.

During the early part of this intertransitional period, the tooth buds for the first and second premolars begin to form between the bifurcations of the primary molars, but they are not ordinarily calcified enough for radiographic visualizations until 4 or more years of age.

The typical intertransitional dentition is usually well aligned and attractive unless perverted by habits or multilated by dental disease. The proper arrangement of the primary dentition does not necessarily guarantee the same for the succedaneous teeth, but most authorities agree that adequate spacing between the primary teeth is conducive to uncrowded situations in the permanent teeth. In some people, the second teeth are so disproportionately large that even significant spacing between the primary teeth fails to achieve an adequate arch length.

First Transitional Period

The first exchange of teeth begins around six years of age and is usually complete within two years. During this time span, the permanent first molars erupt posterior to the primary teeth (sometimes unknown to child or parent) and the more obvious exchange of eight incisors occurs. The staggered exfoliation of the primary incisors and the intermittent periods without a fully erupted complement of teeth, combined with a relative tooth–face size discrepancy at this age, makes for an unaesthetic period in the life of the child.

Another feature of dental development that is deserving of the strongest emphasis is the position of permanent teeth in the jaws prior to their eruption. Not only are the permanent incisors larger than the primary incisors (combined mesiodistal diameters of the maxillary incisors are on an average 7 mm wider, while the mandibular incisors are 5 mm wider), but they are nestled lingual to them within the circumference of a smaller circle. This retracted position of the tooth buds requires some special accommodations during development and dictates considerable adjustments during eruption.

First, in order for the anterior tooth buds to fit within the jaws lingual to their antecedents, they must overlap and assume different vertical levels (Fig. 8.9). In the maxilla, particularly, the lateral incisors are situated behind the centrals and cuspids, and the latter teeth, in addition to their labial position and proximity to the centrals, are also located farthest from the occlusal plane. This variance in the

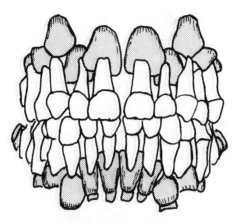

FIGURE 8.9   The positions of the developing succedaneous teeth prior to the loss of any primary teeth.

vertical position of the permanent teeth appears to be a function of the root length; the longer the root, the more removed from the occlusal plane is the tooth bud. Consequently, in the maxilla, the cuspids are very high, followed by the centrals and then the laterals. Contrast this multilevel beginning of the permanent dentition to the single level of the primary dentition. The apical end of the upper central incisors develop just inferior to the nasal aperture, while the cuspids are lateral to the opening. In the lower jaw, the cuspids are so inferior as to be almost at the mandibular border.

The developmental pattern of the anterior teeth forces a vertical and buccal path of eruption. In the maxilla, the tooth buds are already inclined forward so that the labial rotation is not so pronounced as that in the lower incisor, where follicle development is either vertical or even lingually canted.

The labial movement of the anterior teeth follows the pathway established by the gubernaculum and effects an oblique resorption of the roots of the primary teeth (Fig. 8.10). Only in malpositions of the permanent tooth are frontal attacks on the apex seen. Premature loss of an upper primary incisor as a result of trauma can interrupt the pathway of eruption and cause a late emergence of the permanent successor, usually with an excessively labial position.

The upper and lower first permanent molars display contrasting pathways of eruption. Situated in a broader base than its maxillary counterparts, the tooth buds of the lower first permanent molars are mesially and lingually inclined. This position is necessary for development to occur within the curved junction of the ramus and alveo-

FIGURE 8.10 A lateral view of permanent incisor eruption, showing the different path taken by the upper as compared to the lower incisors.

lar process. The eruption path consequently exhibits mesial and lingual arcs (Fig. 8.11). It has been shown that in any one individual, the lower second primary molars and the first and second permanent molars erupt at an identical distance from the posterior border of the ramus.

The upper permanent first molar buds develop with a buccal and distal orientation, a position that allows for the most conservative maxillary tuberosity. As a result, arcial movement is manifested—in this case, distally and buccally.

The relationship between the permanent molars mesiodistally is determined by the vertical alignment of the distal surfaces of the second primary molars. In the situation where a mesial step exists in the primary dentition, the permanent molars assume immediately the interdigitation characteristic of the normal class I molar relationship. No further adjustments are necesary.

If the primary dentition manifests a "flush" terminal plane, the permanent molars are forced into an "end on" relationship unless other adjustments occur. In Baume's type II closed dentition, there is no spacing available anteriorly for mesial migration of the lower teeth. In this situation, the permanent molars remain end on until exfoliation of the second primary molars. At this time, the disparity

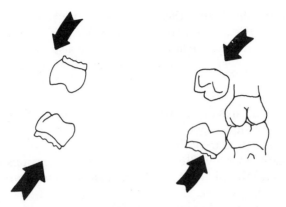

FIGURE 8.11 (*Left*) The paths of eruption of the permanent molars when viewed from the mesial. (*Right*) The paths of eruption of the same molars when viewed from the side.

in size between the primary molar and its replacement, the second bicuspid, might allow some mesial shift of the permanent molar forward, achieving a class I intercuspation. This space has been given the name *leeway space* by Nance, and the process is described as a late mesial shift.

Two other possibilities exist for achieving proper molar relationships in type II individuals: (1) mesial migration of all the lower teeth and (2) a forward adjustment of the condyles in their fossae. Both explanations have been discounted.

In type I dentitions (spaced) with flush terminal planes, early mesial shifts can establish normal occlusal relationships. The spaces between the posterior primary teeth can be closed by the pressure from the erupting permanent first molars. Often this space is adequate to relieve the end on predicament of the accessional teeth.

The upper and lower molars occlude eventually in a manner determined by the terminal plane of the second primary molars, spacing in the primary dentition, the leeway space, cusp–fossa relationships, and muscular forces. This pattern of eruption is also typical of the second and third permanent molars.

Within a few months of the appearance of the first permanent molars, the lower central incisors erupt. The upper centrals emerge a few months later, followed in turn by the lower lateral incisors. The upper laterals are the last teeth to appear in this first transitional phase.

Because of the discrepancy in the mesiodistal crown widths between the primary and permanent incisors, the space available for the permanent teeth after the exfoliation of the antecedents is barely sufficient. Mayne has coined the term *incisor liability* to describe this situation. In order for the larger teeth to fit in a space occupied by the smaller primary teeth, one or more of the following conditions must prevail:

1. Interdental spacing of the primary incisor teeth should exist. This condition suggests the likelihood of extra arch length between the primary cuspids.
2. Intercanine arch width growth should occur. Apposition on the labial alveolar plate allows the canines to move apart only 3 or 4 mm on the average.
3. Intercanine arch length increase through anterior positioning of the permanent incisor teeth. Although the incisors begin development jammed within the alveolar processes, their buccal eruption path eventuates a circumference with the radius greater than the original primary dentition. Baume has reported that the average arch position of the maxillary centrals is 2.2 mm anterior to the primary centrals, a distance that could add up to 3 mm of intercanine dimension.
4. Favorable size ratio between the primary and permanent teeth.

It is not uncommon for the measures to be inadequate in the lower arch, with the result that the incisors are often crowded to varying degrees. If the crowding is not severe, they can unravel as the primary cuspids are lost. This is possible because the combined mesiodistal width of the permanent cuspid and first and second premolar is less than that of the primary cuspid and molars.

Often there is a diastema between the maxillary central incisors, indicating some excess of space.

Thus, at the end of the first transitional stage, the incisors are present, sometimes slightly crowded in the mandible and spaced in the maxilla, with more labial inclination than their antecedents, and the first permanent molars are erupted, usually with an end on relationship.

Second Intertransitional Period

This time period is sometimes referred to as the mixed dentition period, although any coexistence of permanent and primary teeth can theoretically be described as "mixed." It lasts from the full eruption of the incisors until the posterior teeth begin to exchange, a period of approximately 1½ years.

During this time, the vertical dimension of the face is increasing, allowing for a heightening of the alveolar ridges, a process that accommodates continual root growth of the cuspids and premolars. Posteriorly, maxillary tuberosity and mandibular ramal activity prepare room for the second permanent molars.

The maxillary cuspids are still developing lateral to the nose, while the roots of the lower cuspids lie close to the mandibular border. The premolars are nestled in the bifurcations of their antecedents, and there is evidence of resorption on the distal roots.

The posterior permanent succedaneous teeth develop to the distal of the primary teeth they replace, particularly in the lower arch. In part, this posterior placement is made possible by the disparity in size between the crowns of the primary teeth and their successors. On an average, the combined width of the primary cuspid and first and second molars is 1.7 mm greater than the same dimension of the permanent cuspid and premolars (Fig. 8.12). The upper replacement teeth are only 0.9 mm smaller than the primary teeth.

This extra mandibular arch length has been termed as leeway space, implying that this space is allotted for the mesial shift of the lower first permanent molar, which establishes a class I relationship with the upper. In all likelihood, some of this leeway space is taken by the distal movement of the cuspids and premolars, compensating for the incisor liability.

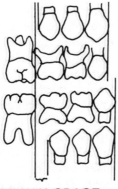

LEEWAY SPACE

FIGURE 8.12 The leeway space, particularly notable in the mandible, which results from the disparity in size between the posterior primary teeth and their successors. Adapted from T.M. Graber: Orthodontics: Principles and Practice. 2nd ed. Philadelphia, J.B. Saunders, 1966.

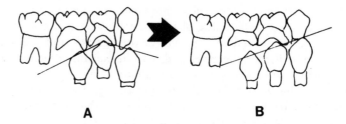

A　　　　　　　　　　**B**

FIGURE 8.13　(A) In the early stage of eruption, the cuspid and first and
second bicuspids present a gable effect whereby the first bicuspid
lies closer to the occlusal plane than the others. (B) With time, the
cuspid normally spurts ahead and is the first of the three to erupt.

If the tips of the cusps of the mandibular cuspids and premolars
are connected, a gable effect is noted (Fig. 8.13A), with the first bicuspid enjoying the peak position. Normally, the sequence of eruption in the lower jaw is cuspid, first premolar, and then second
premolar. In order to achieve such a sequence, late-starting cuspids
must play "catch up" and pass the first premolar as in Fig. 8.13B.
Often these two teeth erupt almost simultaneously, and occasionally
the first premolar is first.

Second Transitional Period

During this active period, the last exchanging of teeth occurs: the primary cuspids and molars are shed, the permanent cuspid and premolars erupt, and the second permanent molars emerge. This
turnover transpires between the ages ten to twelve, but as in all tooth
eruption, chronologic variation is commonplace.

There are several possible sequences of eruption for each arch.
In the maxilla, the most common are (1) first premolar, second premolar, and cuspid and (2) first premolar, cuspid, second premolar.
Occasionally, the cuspid and the second premolar will erupt simultaneously. In the mandible, the order is (1) cuspid, first premolar,
second premolar and (2) first premolar, cuspid, second premolar.
This latter sequence can sometimes be very undesirable, a point to
be discussed later. The mandibular cuspid and first premolar erupt
simultaneously on occasions. The second permanent molars rarely
erupt before cuspids or first bicuspids, but frequently along with
second bicuspids. Preferably, their emergence should follow all the
anterior exchanges, since they add considerably to the mesial component of force that helps drive the first molars into the leeway
space.

The eruption of the upper cuspids often abets the unaesthetic appearance of this stage of the mixed dentition. The clinical picture of such a child presents in the maxilla large (relative) central incisors with a diastema between, flanked by flared, labially inclined laterals. Hidden within the jaw is the source of this temporary misalignment (Fig. 8.14).

The cuspids are erupting labially and mesially, "riding" down the roots of the lateral incisors and displacing them toward the midline. As the cuspid moves nearer occlusion, the force shifts to the other side of the pivot point along the roots of the lateral. As a result, the crown is displaced mesially against the central. Unless there is gingival fibrous tissue trapped in the intermaxillary suture, the wedging action of the cuspids eliminates the diastema between the centrals and also straightens the lateral incisors.

An interesting aspect of cuspid eruption is the differential response of the roots of the permanent lateral incisors as compared to those of the primary cuspids. The eruptive force of the cuspids exerts as much pressure on the laterals as on the primary cuspids, but normally only roots of the latter resorb. The lateral incisors seldom are damaged by the cuspids.

While there is no "best" order for teeth to erupt, an inappropriate eruption sequence for the space available can result in a malocclusion. In many children where mandibular space is at a premium, the sequence cuspid, first premolar, second premolar concentrated within a short time span maximizes the possibility for good alignment. To see how the sequence and timing interrelate, refer to Fig. 8.15. Note that the primary cuspid cannot possibly provide enough space for the permanent cuspid, and the latter tooth must encroach upon the space represented by the first primary

PRESSURE AGAINST PERMANENT LATERAL

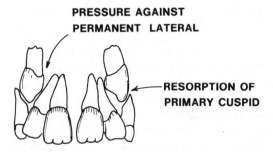

RESORPTION OF PRIMARY CUSPID

FIGURE 8.14   The normal result of pressure from the erupting cuspids is the displacement of the permanent laterals and the resorption of the primary teeth.

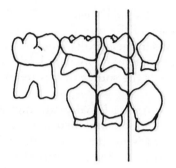

FIGURE 8.15   The vertical lines show that the erupting permanent cuspid must encroach upon the space of the exfoliating first primary molar in order to find room in the arch. In turn, the first bicuspid must utilize some of the space for the second primary molar. Fortunately the second bicuspid is smaller than the second primary molar.

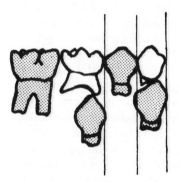

FIGURE 8.16   The crowding and loss of space that can result if the first bicuspid erupts prematurely. The permanent cuspid is crowded out, and the space it needs is lost as the first permanent molar slips forward when the second primary molar is shed.

molar. Ideally, then, the eruption of the first bicuspid is so orchestrated with that of the cuspid that as the latter emerges into the oral cavity the first primary molar is ready to be exfoliated. The perfectly timed loss would allow the mesiodistal width of the larger permanent cuspid to be accomodated in the arch.

At this point, you probably realize that the teeth are "robbing Peter to pay Paul." The first bicuspid is now shortchanged on space, and the only way out of the dilemma is for the second bicuspid to instigate the exfoliation of the second primary molar as the first premolar erupts. The second premolar would now be in jeopardy, just as the cuspid and first premolar before it, if it were not for the excess size of the second primary molar, which allows the successor to slip in.

Consider what could happen in a similarly tight arch if the sequence were altered (Fig. 8.16). If the first premolar erupted ahead of the cuspid and failed to utilize some of the space of the second primary molar, then it would be in contact with the latter tooth positioned too far mesially. The cuspid, without adequate space, would most likely erupt labially. The second premolar would now erupt into the space of the second primary molar, and the leeway space would be consumed by the mesial shift of the posterior teeth. Not all arches are so needful of space conservation, and virtually any sequence and timing seems to work out.

In the transition from the mixed to a complete permanent dentition, the arch length as measured from the mesial of one first permanent molar to its antimere diminishes, while the entire arch circumference increases by virtue of second molar eruption.

Adult Dentition

After the exfoliation of the last primary tooth, the dentition is considered adult by many clinicians, but actually, not until around the twentieth year, when the third molars have erupted and finished root development, is the dentition complete.

Adolescence is characterized by craniofacial growth that puts the finishing touches on the face, so to speak. The nose and chin becomes more prominent, the latter aided by bone deposition. These processes tend to reduce the convex profile of the lip region of children. Additionally, the brow area becomes larger as a result of pneumatization of the frontal sinuses and apposition on the glabella.

The growth of the jaws continues during this period, developing room for the third molars. In many cases the growth is inadequate, and these last molars become impacted. Not infrequently, third molar buds are congenitally missing.

The marked mesial inclination of the posterior permanent teeth diminishes somewhat as the mandible completes its growth from under the maxilla, and the lower incisors tend to become more upright. Often, some crowding of the lower incisors is a result.

The successful attainment of an aesthetic and functional adult dentition depends upon a disease-free or well-maintained mouth, no major trauma, and good genes.

BIBLIOGRAPHY

Baume LJ: Physiological tooth migration and its significance for the development of occlusion. Part I. The biogenetic course of the deciduous dentition. J Dent Res 29:123, 1950

Baume LJ: Physiological tooth migration and its significance for the development of occlusion. Part II. The biogenesis of the accessional dentition. J Dent Res 29:331, 1950

Baume LJ: Physiological tooth migration and its significance for the development of occlusion. Part III. The biogenesis of the successional dentition. J Dent Res 29:338, 1950

Grant D, Bernick S: Formation of the periodontal ligament. J Periodontol 43:17, 1972

Hodson JJ: The gubernaculum dentis. Dent Pract 21:423, 1971

Kronfeld R: The resorption of the roots of deciduous teeth. Dent Cosmos 74:103, 1932

Linden FPGM van der: Theoretical and practical aspects of crowding in the human dentition. J Am Dent Assoc 89:139, 1974

Linden FPGM van der, Duterloo HS: Development of the Human Dentition. Hagerstown, Harper and Row, 1976

Mayne WR: Serial extraction—orthodontics at the crossroads. Dent Clin North Am p 341, July 1968

Mjor IA, Pindborg JJ: Histology of the Human Tooth. Copenhagen, Munksgaard, 1973

Moorrees CFA: The Dentition of the Growing Child. Cambridge, Massachusetts, Harvard University Press, 1959

Nance HN: The limitations of orthodontic diagnosis and treatment. Am J Orthod Oral Surg 33:177, 1947

Provenza DV: Fundamentals of Oral Histology and Embryology. Philadelphia, Lippincott, 1972

Riedel RA: Post-pubertal occlusal changes. In McNamara JA Jr (ed): The Biology of Occlusal Development. Ann Arbor, Michigan, University of Michigan, 1977

Scott JH: The development and function of the dental follicle. Brit Dent J 85:193, 1948

Shulman J: Causes and Mechanisms of tooth eruption—A literature review. J West Soc Periodont 24:162, 1976–1977

Sicher H: Oral Anatomy, St. Louis, CV Mosby, 1952

· 9 ·

Principles of
Cephalometric Analysis

There is something in a face,
An air, and a peculiar grace,
Which boldest painters cannot trace.
William Somerville, The Lucky Hit

For the diagnosis of malocclusion, for treatment planning, and for
the monitoring of appliance therapy, the cephalometric head film
has become an invaluable clinical adjunct. But until the film is traced
and carefully selected landmarks recorded, only general impressions
are afforded. It is when a system of angles, planes, and arcs is exacted
on the tracing that a more precise and meaningful reservoir of information is tapped.

There are almost as many cephalometric systems as there are
investigators. The analyses range from the very simple to the ultra-
complex. Not even a large volume devoted solely to this field could
hope to cover them all adequately. For this reason, the two systems
discussed here have been chosen as representatives of their type and
because they also provide additional insight into craniofacial growth.

There are several classes of cephalometric analyses, but two
have enjoyed more acceptance than the others. The first class in-

cludes those analyses that deal in the quantitation of selected angles and dimensions, the values for which are based either on statistical methods or clinical judgments. Those systems composing the second class are less concerned with absolute values but more with intrinsic symmetry and the proportions of parts.

The majority of cephalometric analyses are of the first type, the values for the angles and linear dimensions representing the means of data collected from an appropriate sampling of the population. Occasionally, the angles and linear measurements are rather arbitrarily assigned, satisfying a clinician's view of what is aesthetically pleasing.

To use one of these analyses, values for specific angles and dimensions are procured from a tracing of a subject and compared to the "standard" values tendered by that analysis. This approach discloses where an individual deviates from the norms established for his or her peer group; it does not always, however, inform about the discrepancy that causes the variation from the norm. For instance, an angle may inform the clinician that the patient in question has a retruded mandible, but it will not disclose if the cause is a short corpus, a small ramus, or a posteriorly positioned glenoid fossa.

To circumvent the inadequacies of some of the statistical approaches, other investigators have designed analyses that deal with proportions and the harmony of parts. These systems attempt to uncover the imbalances within one individual for which compensations have or have not occurred. Analyses of the first type have been developed by Downs, Steiner, Tweed, and countless others; Sassouni, Coben, and Enlow, among many, have contributed to the second type.

CEPHALOMETRIC ANALYSIS USING NORMS

As an example of the type of cephalometric analysis that sets values for the angles and dimensions that purportedly exist in the "average" or "ideal" face, a modification of the Steiner analysis will be discussed. The inclusion of this system of craniofacial appraisal in this book does not imply that it is superior to any other. It was selected because it is relatively simple, because it has been adapted to analyze longitudinal growth, and because it evaluates the three kinds of relationships in the head that are of concern to the clinician.

In Fig. 9.1, the principal landmarks are identified, the planes drawn, and the angles and linear dimensions given their average values. Not all the measurements of the original Steiner analysis are used; the Y axis is drawn from Downs and the IMPA derived from

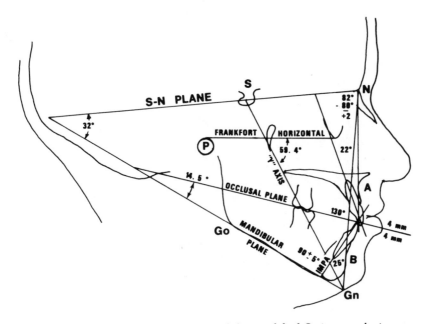

FIGURE 9.1 The planes and angles of the modified Steiner analysis.

Tweed. A table for recording and comparing values from a patient is illustrated in Fig. 9.2.

This modified Steiner analysis fulfills the minimum requirements of an analysis of the first type, having the ability to evaluate the three following relationships:

1. Basal bones to cranium—These measurements relate the positions of the maxilla and mandible to the cranium and to each other.
2. Teeth to basal bones—The positions of the teeth are analyzed with respect to their supporting structures.
3. Teeth to teeth—The interrelationship of the teeth are determined.

Basal Bone Analysis

The first of these relationships is analyzed by five angles, namely, S–N–A, S–N–B, A–N–B, S–N–Go–Gn, and the intersection of the Y axis with the Frankfort horizontal. Since the mandible is more labile than the maxilla, it is most appropriate that the majority of these angles relate to the position and growth tendencies of the lower jaw.

To assay the positions of the components of the face with re-

		NORMAL AVERAGES	THIS ANALYSIS
1.	SNA	82°	
2.	SNB	80°	
3.	ANB	2°	
4.	Y Growth Axis	59.4°	
5.	OCCL to SN	14.5°	
6.	GOGN to SN	32°	
7.	1 to NA	22°	
8.	1 to NA mm	4 mm	
9.	1 to NB°	25°	
10.	1 to NB mm	4 mm	
11.	1 to 1	130°	
12.	IMPA	90° ± 5°	

FIGURE 9.2   Summary sheet for patient evaluation.

spect to the cranium, Steiner selected the plane from sella to nasion as a frame of reference. The points that establish this plane were chosen because they represent the anterior cranial base, they are midsagittal structures, and they are roentgenographically easy to locate.

S–N–A, S–N–B, and Tweed's Growth Classification

To evaluate the anteroposterior position of the maxilla and mandible, Steiner borrowed from Reidel the angles S–N–A and S–N–B.

S–N–A is the angle established by the intersection of the plane S–N with a line drawn from N to A extended through the occlusal plane. Since point A is on the maxilla, the angle S–N–A reflects the anteroposterior position of the maxilla as it relates to the base of the skull. Steiner judged through clinical experience and observation of aesthetically acceptable profiles that the angle S–N–A should average 82° in the adult.

Since the middle face is so intimately involved with the anterior cranial base, the angle S–N–A is not expected to change significantly during growth. This is not true of the angle S–N–B.

S–N–B is the angle produced by the intersection of the plane S–N with the line N to B, and it reflects the anteroposterior position of the mandible as it relates to the cranial base. Since, as discussed in Chapters 4 and 5, the mandible is retruded in early life and achieves near parity with the maxilla by demonstrating a longer span of growth, the angle S–N–B can be expected to increase some 4 to 6° during facial development. According to Steiner, an angle of 80° is ideal for an adult.

The angle A–N–B has become an important clinical determinant, since it discloses the relationship of the maxilla to the mandible, irrespective of the cant of the cranial base. This angle can be expected to close if a favorable pattern of growth is displayed by the child. An adult ANB angle of 2° is considered ideal.

Tweed adapted the change in angle A–N–B as a simple clinical test for determining the type of growth manifested by a child. This analysis is predicated on the belief derived from clinical experience that, except for a small percentage of patients, orthodontic therapy is aided by a "good" growth pattern. This term is defined as the condition where mandibular growth has a strong forward component to it.

To determine the pattern of growth, Tweed advocated taking serial roentgenograms 15 to 18 months apart and superimposing the tracings on S–N with S as the reference point. Based on the possible changes in A–N–B, the following classification was created:

Type A Growth Trend

The middle and lower face of the patients grow forward at the same rate and consequently the angle A–N–B remains constant. According to Tweed, 25 percent of his patients fell into this category. Treatment on children with such a growth pattern can be difficult.

Type B Growth Trend

In these children, the middle face grows forward with the cranial base, but the mandible displays a more vertical than horizontal component. Consequently, A–N–B increases and the profile retrudes from the ideal established by Steiner and others. Tweed found that about 15 percent of his patients fell into this category. Orthodontic treatment is limited by this growth pattern; in some cases, the handicaps are formidable.

Type C Growth Trend

The majority of patients (60 percent) exhibit a reduction in the angle A–N–B during growth. Only a small percentage of this group exceeds the norm and becomes prognathic. Orthodontists appreciate the help from nature during the treatment of type C individuals.

The Y Axis

Another useful parameter to evaluate the position of the mandible and the course of its growth is the angle the Y growth axis makes with the Frankfort horizontal. This angle is formed by the intersection of the line S–Gn and the Frankfort horizontal, which connects orbitale and porion. Downs found a mean of 59.4° among 20 subjects with an age range of 12 to 17 years.

The Y growth axis, as it is sometimes called, essentially establishes the direction of growth of the mandible; an angle more acute than average tends to denote a type C grower, and an angle more obtuse suggests a type B. The Y axis is not immutable, however, and the angle it makes with the Frankfort horizontal can close or open, but never more than a few degrees.

The Mandibular Plane

Additional information about the growth tendency of the mandible is achieved by measuring the angle formed by the intersection of the mandibular plane (Go–Gn) with S–N. The mandibular plane is drawn along the inferior border of the mandible and is extended in a posterior direction to form an angle with the plane S–N. Steiner found a favorable GoGn–SN angle in a balanced face to be 32°.

Larger angles are often associated with downward growing mandibles, retruded chins, and long faces. Called *high angle* cases by orthodontists, they tax the therapist who is trying to correct a malocclusion while not accentuating these unfavorable features.

Teeth to Bone Analysis

The position of the incisors in the face is accomplished by linear and angular measurements. The angulation of the incisors with respect to basal bones is achieved by determining the angles formed by the intersection of the long axes of the upper ($\underline{1}$) and lower ($\overline{1}$) incisors with N–A and N–B, respectively. Since a perfectly angulated incisor can be positioned too far forward or too far back, the distances from the $\underline{1}$ to Na and $\overline{1}$ to N–B, measured from the incisal edges along the occlusal plane, are taken into account. According to Steiner, the maxillary central incisor should assume a 22° angle with N–A and be 4 mm ahead of it. The lower incisors should be situated at a 25° angle to NB and 4 mm anterior to it.

Additional information about the position of the lower incisor is gained through the utilization of the incisor–mandibular plane angle (IMPA) of Tweed. The average 90° angle (± 5°) was recommended, the value being based upon what he felt contributed to stability and aesthetics.

Teeth to Teeth Analysis

The interdental relationship is achieved by measuring the angle formed by the intersection of the long axes of the central incisors. In an "average" face, 130° constitutes a favorable angle.

An analysis as just described is usually based on standards derived from a population of well-balanced faces. When a disharmony occurs in a face, it must be realized that positive compensations can occur. For instance, if the basal bones are not properly aligned anteroposteriorly, the incisors can still assume balanced and functional positions. Although the angular values in such a face may differ from the standards, the whole of the face still works and no malocclusion exists. Experience is often very helpful in identifying those deviations that lie even beyond the bounds permitted by compensated faces.

The values given in these analyses are often expressed as means plus or minus standard deviations or averages with ranges of normal. It is the obligation of the clinician to judge whether the deviations from the central tendency lie within or beyond the limits of acceptability. Most of us vary from the standards to lesser or greater degrees, yet, in the main, our appearances are quite tolerable. Thus, norms are not rigid values; they are guidelines to be tempered with clinical judgment.

CEPHALOMETRIC ANALYSIS OF INTRINSIC CRANIOFACIAL FORM

The Counterpart Principle

The type of analysis previously discussed is excellent for comparing a subject under question to the norm established from a population of peers. It does not, however, provide much insight into the root causes of angular deviations. For instance, an A–N–B difference of 5° in a 12-year-old girl forebodes a mandibular deficiency and a convex profile. The Y axis and the mandibular plane help to clarify the position and growth tendency of the lower jaw. What is not revealed, unfortunately, is the disharmony between parts for which no compensation has occurred.

It is possible with more parameters included in an analysis to better distinguish the underlying basis for the abnormal growth pattern. Yet it remains difficult to ascertain the discrepancies within an individual by comparing him to a population. To "individualize" the appraisal of a craniofacial complex, other researchers and clinicians have turned to a variety of analyses that deal in proportions or intrinsic form.

One such system, by Enlow, has evolved into a general approach for describing craniofacial growth as well as uncovering disproportionate growth patterns. The analysis is built around the concept that certain regions of the head have counterparts or other regions that are related to and balance with them.

We have already seen a little of this thinking involved in the description of horizontal growth anterior and posterior to Ricketts' PTV line. In particular, the relationship between the anterior cranial base, maxilla, and chin point was discussed. Enlow has developed this approach to the fullest, and, unfortunately, we can only hope to present an overview of a complex analysis in this text.

The PM Plane

Enlow divides the face into anterior and posterior segments by a vertical line that is similar to, but not identical to, the PTV line. This line (Fig. 9.3) drops from the junction of the anterior and middle cranial fossa (and the inferior junction between the frontal and temporal lobes of the brain), perpendicular to the neutral line of the orbit, passing almost exactly along the posterior surface of the maxillary tuberosity, and through the posterior boundary of the corpus, the lingual tuberosity.

This line constitutes the PM (posterior maxillary) plane, a natural boundary that separates key sites of growth and remodeling. The superiormost point (junction of the anterior and middle cranial fossae) was chosen because Enlow considers that the floor of the cra-

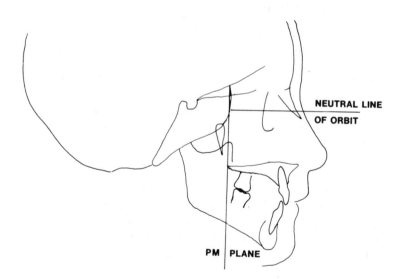

FIGURE 9.3 The PM plane (posterior maxillary), which drops from the junction between the anterior and middle cranial fossae, perpendicular to the neutral line of the orbit, touching the posterior maxillary tuberosity and the mandibular lingual tuberosity.

nium has developed in phylogenetic association with the brain and that the brain has a natural boundary between lobes.

Because of the special association of the senses sight and smell to the brain, the nasomaxillary complex is especially molded beneath the anterior cranial fossa. The direction of the visual sense is perpendicular to the new PM plane; the maxillary tuberosity is perpendicular to the orbit axis and so situated that the PM plane passes almost exactly along its posterior surface. The relationship between the structures is so interwoven as to preclude happenstance.

The lingual tuberosity is the mandibular equivalent to the maxillary tuberosity. It constitutes the functional boundary between the corpus and ramus of the mandible. Although morphologically the mandible is one bone, the posterior part (ramus) and the anterior part (corpus) are related to different fields of the head.

The Counterparts

The PM plane, much like the PTV, divides the face into two vertical areas and facilitates the discussion of the relationships between the components of these areas. For instance, the face anterior to the PM plane exhibits a vertical array of structures that are all counterparts of each other—they all normally fit each other. Superiorly, the frontal lobe of the brain sets the stage. Beneath are the anterior cranial fossa, the ethmomaxillary complex, the palate, and the maxillary arch, all with posterior boundaries along the PM plane. These structures are all intimately related by sutures. But the corpus of the mandible, although unattached by bony tissue, is nevertheless, a counterpart to all of them. Thus, for "normal" development, the anterior cranial fossa, the maxillary complex, and the corpus should theoretically match.

The posterior vertical array of counterparts are the temporal lobe, the middle cranial fossa, the posterior oropharyngeal space, and the ramus of the mandible.

With this division and these counterparts in mind, we can gain yet another insight into normal and abnormal craniofacial growth.

DIMENSIONAL BALANCE

Probably the most important element in counterpart analysis is the cranial base. Two aspects of the cranial base profoundly affect the positions of the structures related to it. They are (1) its growth and (2) its flexure. We will discuss each in turn. In Fig. 9.4, note that brain growth in this area helps create new bony support beneath it. The sphenoid is displaced upward and forward, accomodating temporal brain expansion, while the anterior cranial fossa is displaced forward in response to brain growth there. In addition, the increase

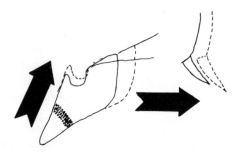

FIGURE 9.4 The encroachment of the airway by the enlarging sphenoid dictates that the anterior cranial base must grow forward, carrying the middle face with it.

in size of the sphenoidal sinus by drift adds to the overall size of the bone. One consequence of this growth of the sphenoid is an encroachment into the nasal and oropharyngeal spaces, a condition that cannot be tolerated. As a result, both the maxilla and mandible move anteriorly and inferiorly to maintain the patency of these cavities.

The maxilla, because of its intimate association with the anterior cranial fossa and because the nasal septum is buttressed against the sphenoid, tends to be displaced along with the frontal lobe of the brain. As a consequence, the PM plane is positioned forward in space and the posterior boundary of the maxillary tuberosity still coincides with it.

The position of the maxilla, then, has been influenced by the growth of the cranial base anterior to the spheno-occipital synchondrosis. The mandible is located anatomically on the opposite side of the synchondrosis and is therefore not influenced to the same degree by cranial base growth. This means that the mandibular counterpart of the middle cranial fossa and oropharyngeal space, the ramus, must bridge the expansion of these areas through its own initiatives. Thus, condylar growth, posterior ramal apposition, and complicated V principle resorption on the leading edge of the ramus maintain the lingual tuberosity on the PM plane.

Similarly, the corpus of the mandible must pace its growth with the counterparts above it, the anterior cranial fossa and maxillary complex. Optimally, the corpus increases its horizontal dimensions by ramal erosion sufficient to allow eruption of posterior teeth and to match the posterior growth and anterior displacement of the maxilla. Bear in mind that the corpus is not structurally related to the cranial base in the same way as the maxillary complex, so as a counterpart, its relationship is most tenuous. The ramus, a posterior counterpart, by virtue of its growth and size, greatly influences the anterior counterpart, the corpus.

Thus far, the analysis of the balance and harmony between the counterparts has been limited to a comparison of their horizontal dimensions. It has been emphasized, for instance, that the depth of the anterior cranial base must be matched by the depth of the maxilla and corpus of the mandible. But the lengths of these structures are only part of the relationships. Also involved in the creation of a human profile is the alignment of the counterparts.

ALIGNMENT

Compare the two tracings in Fig. 9.5. All the respective components are identical in dimension. The only purposeful change in the drawings is a decrease in the flexure of the cranial base. As a consequence of this enlarged angle, spatial changes or alignment changes are made in other structures. The mandible now assumes a retruded position behind the PM line, with the lingual tuberosity becoming posteriorly situated. The anterior cranial base is now lower in space, and the maxilla is carried with it. The mandible, already at a deficit because of the increase in overall horizontal cranial base dimension, is further retruded by a backward rotation in response to this added vertical component. Plainly, then, alignment of the counterparts has

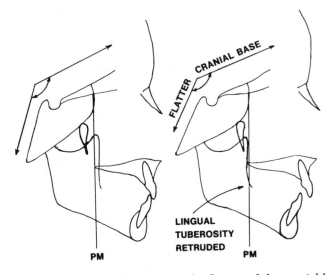

FIGURE 9.5   The relationship between the flexure of the cranial base and the retrusion of the mandible.

as much to do with proportions as do dimensions. In most humans, the alignment and the dimensions of the counterparts are not perfect, but there are enough compensations made that most of us look reasonably good. Where cranial base flexure, for instance, may raise havoc with alignment, increased or decreased dimension of a part may effect a balanced face.

Analysis of Growth by the Counterpart Principle

To explore this fascinating relationship in "normal" compensated states and in uncompensated malocclusions, it might be easier to turn to stick diagrams of the human face as devised by Enlow. Figure 9.6 shows the positions of the lines over a tracing of a head film. The points were selected to best reflect the junctions of the cranial floor with the condyle of the mandible and with the nasomaxillary complex. (For full details, see Enlow.) The lines denote the critical areas of the cranial base, circumscribe the maxillary complex, and depict the mandible in space. Note particularly the alignment of the parts to the PM plane.

With this technique, note how the changes in Fig. 9.5 can be easily and informatively presented (Fig. 9.7). One can see clearly

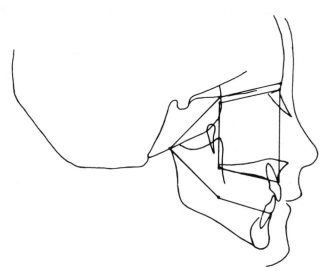

FIGURE 9.6 The essential dimensions and relations of the cranial base, middle face, and mandible as depicted by lines. Adapted from D.H. Enlow: Handbook of Facial Growth. Philadelphia, J.B. Saunders, 1975.

how the straightening of the cranial base has thrust the maxilla forward and downward in space. As a result, the mandible has rotated at the condyle, compounding the already posterior position of the lingual tuberosity. The counterparts of the face are no longer aligned on the PM plane—at least with dimensions as they are. When this situation remains uncompensated, a classic class II malocclusion results—the mandible is retruded and the molars exhibit a class II relationship.

But, as stated before, most of us are walking compensations. For instance, in the previous example characterized by a forward-inclined cranial base, dimensional changes can adjust for alignment imbalances (Fig. 9.8), but without them, the mandible will remain retruded both at the PM plane and anteriorly. One consequence of a forward-inclined cranial base is an increase in the horizontal dimension of the middle cranial fossa. The counterpart to this region is the ramus, and if it can establish a new horizontal dimension, then realignment becomes a possibility. Note in the stick diagram how ramal growth could effect a repositioning of the corpus and return the profile to normal (shaded area). To see how this looks in real people, compare the two tracings in Fig. 9.9. The one on the left is an uncompensated class II malocclusion; the one on the right shows how an increase in ramal width has realigned the counterparts.

The reverse phenomenon can occur in that altered alignment can compensate for discrepancy in the lengths of counterparts. In Fig. 9.10, observe that a long corpus does not necessarily cause a prognathic profile if the alignment along the PM plane has been modified. The flattened cranial base normally precipitates a class II profile, but in this individual, the long corpus has adjusted for the lingual tuberosity being located posterior to the PM plane and for the rotation backward of the mandible. The molar relationship is a class II type, despite a mandible shaped as those seen in class III malocclusions.

While it is beyond the scope of this book to pursue the counterpart principle to extremes, it is worth adding that a backward-inclined cranial base can have exactly opposite effects. The maxilla is elevated and the mandible is thrust forward, rotating upward. The resulting general appearance varies from a flattened profile to one that is outright prognathic.

The counterpart analysis, like the Ricketts' center of least growth, provides a unique way of looking at the overall growth of the face. Its main attraction is its ability to isolate those components of the head which either contribute to harmony or those that perpetrate disharmony.

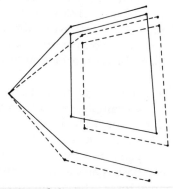

FIGURE 9.7   Line drawing of the head illustrating the influence of a flattened cranial base in maxillary and mandibular positions. Adapted from D.H. Enlow: Handbook of Facial Growth. Philadelphia, J.B. Saunders, 1975.

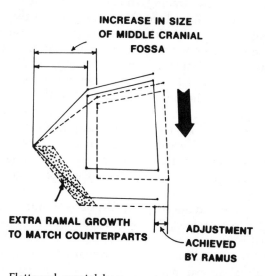

FIGURE 9.8   Flattened cranial base causes an increase in the size of the middle cranial fossa. As a result, the mandible will end up retruded unless extra ramal growth compensates as shown in the line drawing.

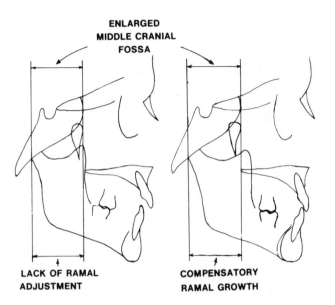

FIGURE 9.9 A comparison of two individuals with flattened cranial bases and enlarged middle cranial fossae. The individual on the left suffers from mandibular retrusion as a result; the one on the right has reestablished an orthognathic profile utilizing compensatory ramal growth.

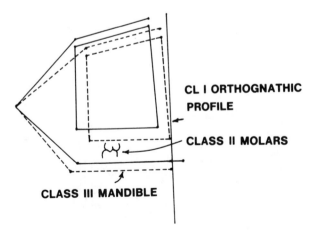

FIGURE 9.10 The unique case of a flattened cranial base leading to a class II molar relationship in a class III mandible. The compensations result in a normal class I profile. Adapted from D.H. Enlow: The Human Face. Philadelphia, J.B. Saunders, 1975.

THE CEPHALOMETRIC HEAD FILM

Most radiographic head films are still taken according to the standards of Holly Broadbent, who established them almost 50 years
ago. His format for taking head films was quickly adopted by the sci-
entific community, not only because he pioneered work in the field,
but because of the consensus that his instrumentation, radiographs,
and data were impeccable. Probably his major contribution was designing a cephalostat (or head holder), which restrained the head in a
precise manner. Adjustable ear rods (mechanical porion), which fit
into the external auditory meati, held the head stable anteroposteriorly and prevented lateral rotations (Fig. 9.11). The source of the
x-ray was set 5 feet from the middle of the cephalostat, and the central beam was designed to pass through the ear rods. Although the
clinical cephalostats used today are much less obtrusive and expen-
sive than Broadbent's research instrument, the positioning of the
head and the distance between the x-ray unit and the head holder
remain the same.

For most analyses, only lateral cephalograms are taken. When
an anteriorposterior view is desired, the source is positioned behind
the head with the central beam at the level of the ear rods. The distance remains 5 feet between the cephalostat and the x-ray head.

The modern cephalostat guarantees that the orientation of the

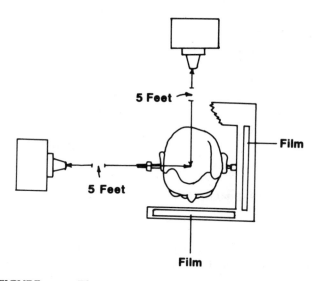

FIGURE 9.11   The orientation of the head in the cephalostat.

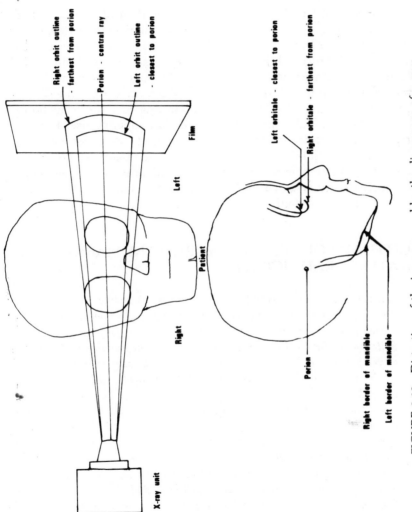

FIGURE 9.12 Distortion of the image caused by the divergence of x-rays.

patient's head is standardized and consistent, allowing valid comparisons between individuals, even when different machines are used. For longitudinal studies, the head holder fosters reproductibility; the recording of growth changes would be futile if head positions were haphazard.

Outside of technical errors associated with head placement and exposure times, one inherent and unavoidable flaw in any cephalometric system is the lack of parallelism of the x-rays. Because only at an infinite distance will x-rays not diverge, users of head films have learned to deal with some distortion in the image. The right side of the head, which is closest to the source and farthest from the film, is somewhat magnified (Fig. 9.12). This means that those anatomic structures with right and left sides seldom superimpose. For this reason, midsaggital landmarks are preferred whenever possible. When faced with bilateral structures, it is often necessary to restrict arbitrarily the tracing to one side; others advocate an average position between the two images. Either method is suitable as long as the procedure is consistent.

GLOSSARY OF CEPHALOMETRIC LANDMARKS (LATERAL RADIOGRAPH)

Anterior nasal spine (ANS)—Pointed process extending from the nasal floor formed by the meeting of both nasal margins at the midline.

Basion (Ba)—Midpoint of the anterior border of the foramen magnum.

Bolton point (BP)—Highest point in the profile roentgenogram at the notches on the posterior end of the occipital condyles on the occipital bone.

Cc–Intersection of the line drawn from Pt (perpendicular to Ba–N) with the plane Ba–N.

Dc—Bisection of the condyle neck as high as visible in the cephalometric film below the fossa.

Eva—Forking of the stress lines at the terminus of the oblique ridge on the medial side of the ramus.

Gnathion (Gn)—Point on the symphysis at the crossing of the Y axis determined by the intersection of the facial and mandibular planes.

Gonion (Go)—Point at the inferior posterior border of the mandible located by bisecting the angle formed by the mandibular plane and a line drawn along the posterior border of the condyle and ramus.

Infradentale—Superior labial termination of the cortical plate just below the mandibular incisors.

Mental protuberance (Pm)—Bone crest located at the superior aspect of the compact bone on the anterior contour of the symphysis.

Nasion (N)—Anterior termination of the suture between the frontal and nasal bones.

Orbitale (O)—Lowest point on the infraorbital margin, usually the left.

Pogonion (Po)—Most anterior point on the midline of the mandible.

Point A (A)—Deepest midline point on the maxilla between the anterior nasal spine and prosthion.

Point B (B)—Deepest midline point on the mandible between infradentale and pogonion.

Porion (P)—Highest point on the hard tissue of the external ear canal. Often determined mechanically using the ear rods of the cephalostat.

Prosthion—Inferior labial termination of the cortical plate just above the maxillary incisors.

Pterygoid point (Pt)—Lower lip of the foramen rotundum.

Pterygoid vertical (PTV)—Line perpendicular to the Frankfort horizontal that lies tangent to the pterygoid fossa.

Registration point (R)—Center of the line perpendicular to N–Bp drawn from S.

Sella turcica (S)—Center of the bony crypt that contains the pituitary.

Xi—Centroid of the ramus.

BIBLIOGRAPHY

Broadbent BH: A new X-ray technique and its application to Orthodontia. Angle Orthod 1:45, 1931.

Coben SE: The integration of facial skeletal variants. Am J Orthod 41:407, 1955

Downs WB: Variations in facial relationships: Their significance in treatment and prognosis. Am J Orthod 34:812, 1948

Enlow DH: Handbook of Facial Growth. Philadelphia, Saunders, 1975

Enlow DH, Azuma M: Functional growth boundaries in the human and mammalian face. Birth Defects 11(7):217, 1975

Enlow DH, Kuroda T, Lewis AB: The morphological and morphogenetic basis for craniofacial form and pattern. Angle Orthod 41:161, 1971

Enlow DH, Kuroda T, Lewis AB: Intrinsic craniofacial compensations. Angle Orthod 41:271, 1971

Enlow DH, Moyers RE, Hunter WS, McNamara JA: A procedure for the analysis of intrinsic facial form and growth. Am J Orthod 54:6, 1969

Sassouni V: A roentgenographic cephalometric analysis of cephalo-facio-dental relationship. Am J Orthod 41:735, 1955

Steiner CC: Cephalometrics for you and me. Am J Orthod 39:729, 1953

Steiner CC: Cephalometrics as a clinical tool. In Kraus BS, Reidel RA (eds): Vistas in Orthodontics. Philadelphia, Lea and Febiger, 1962

Tweed CH: Clinical Orthodontics, Vol 1. St. Louis, Mosby, 1966

Index

Acid mucopolysaccharides. *See*
      Glycosaminoglycans
Adolescence, 159
Alveolar bone, 45, 62, 74, 104, 141,
      146
Alveolar process, 55, 68, 69, 93, 145,
      155
Ameloblasts, 140, 141
A-N-B angle, 163, 165, 167
Anencephaly, 111
ANS-Xi-Po angle, 127
Apposition, 11, 13, 20, 32-33, 34-37,
      42, 53-55, 56, 57, 58, 62,
      69-75, 94, 95, 131, 150
A-Pt-ANS angle, 127
Arc of growth, mandibular, 131-137
Arthritis, 12
Articular cartilage, 12, 15, 17, 19
Articular surfaces, 12, 19

Ba-N-A angle, 90
Ba-Pt-Gn angle, 136-137
Basal bone analysis, 163-166
Basioccipital, 60
Basion, 50

Basion (*Cont.*)
   -nasion distance, 81-82, 85-87,
      89, 90
   superimposition on, 80, 81-82
Baume, L. J., 16, 98, 100, 106, 149,
      152
Bicuspids. *See* Premolars
Bite plane, 108
Bizygomatic growth, 58
Bjork, A., 131
Blood supply, 15
Blood vessels, 22, 104
Bone
   adaptability of, 38-39
   bending and surface curvature
      of, 39-40
   compact, 20, 21, 22
   deformation of, 10, 38
   differentiated from cartilage, 5
   as focal point of research, 2
   formation, methods of, 11-20
   growth
      contradictions of, 38-39
      differential, 14
      direction of, 34-37
      mechanisms of, 5-28, 94, 95.
         See also specific mechanism